Optics

Optics

Volume 2 of *Modern Classical Physics*

KIP S. THORNE *and* **ROGER D. BLANDFORD**

PRINCETON UNIVERSITY PRESS

Princeton and Oxford

Published by Princeton University Press
41 William Street, Princeton, New Jersey 08540
6 Oxford Street, Woodstock, Oxfordshire OX20 1TR

press.princeton.edu

All Rights Reserved
ISBN (pbk.) 978-0-691-20736-0
ISBN (e-book) 978-0-691-21556-3

British Library Cataloging-in-Publication Data is available

Editorial: Ingrid Gnerlich and Arthur Werneck
Production Editorial: Mark Bellis
Text and Cover Design: Wanda España
Production: Jacqueline Poirier
Publicity: Matthew Taylor and Amy Stewart
Copyeditor: Cyd Westmoreland

This book has been composed in MinionPro, Whitney, and Ratio Modern by Windfall Software, Carlisle, Massachusetts, using ZzTEX

Printed on acid-free paper.

Printed in China

10 9 8 7 6 5 4 3 2 1

A NOTE TO READERS
This book is the second in a series of volumes that together comprise a unified work titled *Modern Classical Physics*. Each volume is designed to be read independently of the others and can be used as a textbook in an advanced undergraduate- to graduate-level course on the subject of the title or for self-study. However, as the five volumes are highly complementary to one another, we hope that reading one volume may inspire the reader to investigate others in the series—or the full, unified work—and thereby explore the rich scope of modern classical physics.

To Carolee and Liz

CONTENTS

T2 Track Two; see page xv

BOXES

The word "optics" originally referred to the study of light, but today it has come to mean the study of all types of waves, for example, electromagnetic waves of all frequencies, from gamma rays to radio; gravitational waves; quantum waves of probability amplitude governed by the Schrödinger equation; water waves on the ocean's surface and on lakes, ponds, and rivers; sound waves in solids—seismic waves (compressional, shear, torsion, surface, etc.), flexural waves on a whip or a stiff beam or a skyscraper; and Alfvén waves in a plasma.

Optics focuses largely (but of course not exclusively) on fundamental phenomena and computational techniques that are common to all these waves. For example, in the short-wavelength limit, all waves and their amplitudes, polarizations, and energies are transported by quanta that are conserved (in the absence of interactions) and that travel along rays governed by Hamilton's equations. When the wavelengths are longer, all waves exhibit diffraction and spreading. Whatever their wavelength may be, all waves also exhibit interference. Most waves exhibit nonlinear wave-wave coupling that becomes stronger as the waves strengthen. Most waves also exhibit dispersion (with their dispersion relation playing the role of the hamiltonian in the geometric-optics domain of short wavelengths). And some waves can have dispersion stably balanced by nonlinear self-coupling, creating a nonlinear, solitary wave that holds itself together with unchanging shape as it propagates. Solitary waves can even retain their shapes after passing through one another and interacting nonlinearly.

Optical (wave) phenomena are part of everyday human experience, so much so that every science student should ponder how they come about—for example, rainbows; wiggly caustics of brightness on the bottom of a swimming pool produced by sunlight traveling through choppy water; diffraction patterns on the edges of shadows; the twinkling of stars produced when the starlight passes through the earth's turbulent atmosphere; the steepening and then breaking of ocean water waves as they near shore; the production, by storms at sea, of waves strong enough to challenge surfers; and the merging of colored spots into smooth images as one backs away from a pointillist painting.

When we (the authors) were students, optics was a standard part of physics curricula—often a one-quarter or one-semester undergraduate course. Over the decades since then, optics courses have largely disappeared from physics curricula, moving into engineering courses or disappearing altogether. Whatever the reason, this is no longer justified. Over the past few decades, remarkable advances in optics technology and a growing appreciation of the wide-ranging applicability of optics' central principles have made optics once again highly relevant to a physicist's education, as well as to engineers and other natural scientists.

This transformation began, in part, with the inventions of the laser and of visual-frequency optical elements to measure and manipulate laser light. More recently, wave-wave coupling of light in nonlinear crystals has accelerated the visual-frequency optics revolution. Products of the revolution include holograms for making 3-dimensional images—and perhaps in the not-too-distant future, volume holograms for glasses-free 3-dimensional movies; supermarket scanners; LCD displays for computer and television screens; laser pointers whose green light has been nonlinearly frequency doubled from the infrared; optical fibers that carry densely packed information encoded in light; optical switches and other devices for manipulating that optical information and for optical signal processing; CDs, DVDs, and BDs for storing that information (usually music or movies); mode-locked lasers for producing ultra-sharp pulses of light, which underlie optical-frequency combs (a foundation for better spectrographs, for the most stable clocks of the near future, and for transporting frequency standards between optical and radio frequencies); LIGO's laser interferometer gravitational-wave detectors, which have created the new field of gravitational-wave astronomy; nonlinear-crystal squeezers for manipulating the quantum electrodynamical vacuum—a crucial element of current-generation LIGO; among many, many others.

In the meantime, microwave and millimeter-wave technologies have also advanced remarkably, for example, making possible the microwave communication links that today are ubiquitous around the world and the linking of millimeter-wave telescopes from Greenland to the south pole and from Hawaii to Europe, to produce, via interferometry, a single telescope the size of Earth (the Event Horizon Telescope), which has begun making images of magnetized plasma swirling around black holes at the centers of galaxies.

In view of these optics revolutions, scientists' growing understanding and appreciation of optics phenomena throughout the natural world, and also everyone's everyday experience with intriguing optical devices, we (the authors) and many of our colleagues advocate moving optics back into the standard physics curriculum. We hope that this optics textbook will facilitate that move.

THE CLASSICAL-QUANTUM CONNECTION

Although this book's principal focus is classical optics (i.e., optics in the domain of classical physics), inevitably we make frequent reference to quantum mechanical concepts and phenomena, and we often use quantum concepts and techniques in the classical domain (particularly when dealing with short-wavelength phenomena: geometric optics and nonlinear wave-wave coupling). This is because classical physics arises from quantum physics as an approximation. The roots of classical physics are in the quantum domain, and those quantum roots feed indelible imprints into classical physics. Moreover, fully quantum mechanical aspects of optics are becoming so important for technology that, without apology, we touch on them from time to time in this classical book.

GUIDANCE FOR READERS

The amount and variety of material covered in this book may seem overwhelming. If so, keep in mind that

- *the primary goals of this book* are to teach the fundamental concepts and principles of optics (which are not so extensive that they should overwhelm); to illustrate those concepts and principles in action; and through our illustrations, to give the reader some physical and intuitive understanding of optics.

We do not intend the reader to master our many illustrative applications.

We have aimed this book at advanced undergraduates and first- and second-year graduate students, of whom we expect only (1) a typical physics or engineering student's facility with applied mathematics and (2) a typical undergraduate-level understanding of classical mechanics, electromagnetism, and—for modest portions of the book—quantum mechanics. We also target working scientists and engineers who want to improve their understanding of optics.

Optics in this book comprises four chapters and can be taught in a one-quarter course or even a half semester, though a more leisurely full semester course will leave the student with a broader and deeper understanding of the wide scope of optics. For those readers who would like an even briefer introduction to modern optics, we have labeled as "Track Two" some sections that can be skipped on a first reading, or skipped entirely—but are sufficiently interesting that most readers may choose to browse or study them. Track-Two sections are identified by the symbol T2 . For readers who want more detailed and comprehensive introductions than ours, we offer some recommendations at the end of each chapter.

This book is the second of five volumes that together constitute a single treatise, *Modern Classical Physics* (or "MCP," as we shall call it). The full treatise was published in 2017 as an embarrassingly thick single book (the electronic edition is a good deal lighter). For readers' convenience, we have placed, at the end of this volume, the Table of Contents, Preface, and Acknowledgments of MCP. The five separate textbooks of this decomposition are

- Volume 1: *Statistical Physics*,
- Volume 2: *Optics*,
- Volume 3: *Elasticity and Fluid Dynamics*,
- Volume 4: *Plasma Physics*, and
- Volume 5: *Relativity and Cosmology*.

These individual volumes are much more suitable for human transport and for use in individual courses than their one-volume parent treatise, MCP.

The present volume is enriched by extensive cross-references to the other four volumes—cross-references that elucidate the rich interconnections of various areas of physics.

In this and the other four volumes, we have retained the chapter numbers from MCP, and, for the body of each volume, MCP's pagination. In fact, the body of this volume is identical to the corresponding MCP chapters, aside from corrections of errata (which are tabulated at the MCP website http://press.princeton.edu/titles/MCP .html), and a small amount of updating that has not changed pagination. For readers' cross-referencing convenience, a list of the chapters in each of the five volumes appears immediately after this Preface.

EXERCISES

Exercises are a major component of this volume, as well as of the other four volumes of MCP. The exercises are classified into five types:

1. *Practice*. Exercises that provide practice at mathematical manipulations (e.g., of tensors).

2. *Derivation*. Exercises that fill in details of arguments skipped over in the text.

3. *Example*. Exercises that lead the reader step by step through the details of some important extension or application of the material in the text.

4. *Problem*. Exercises with few, if any, hints, in which the task of figuring out how to set up the calculation and get started on it often is as difficult as doing the calculation itself.

5. *Challenge*. Especially difficult exercises whose solution may require reading other books or articles as a foundation for getting started.

We urge readers to try working many of the exercises—especially the examples, which should be regarded as continuations of the text and which contain many of the most illuminating applications. Exercises that we regard as especially important are designated by **.

In each chapter of this book, we present a set of fundamental optics concepts, principles, and computational tools, and then illustrate them in action with some applications. Naturally, these applications are eclectic and reflect our own interests and experience, but we have striven to choose examples from a wide range of subjects and fields that use optics.

BRIEF OUTLINE OF THIS BOOK

In Chap. 7, we develop and apply *geometric optics* for waves whose wavelengths are much smaller than the radii of curvature of phase fronts and much smaller than the inhomogeneity scales of the medium through which the waves propagate, and whose periods of oscillation are much shorter than the timescales on which the medium is changing. We show, in a very general context, how to derive, from the waves' wave equation, the description of the waves as transported by conserved quanta that travel along rays in a manner described by hamiltonian mechanics, with the waves' dispersion relation playing the role of the hamiltonian. Specializing to nearly planar waves propagating along an optic axis (the paraxial approximation), we develop a formalism that is powerful for practical applications. Among the phenomena we explore with this formalism are caustics, catastrophe theory as a tool for classifying and understanding caustics, practical optical instruments (lenses, telescopes, microscopes, etc.) and aberrations in them, and gravitational lensing of starlight by black holes and galaxies.

In Chap. 8, we study *diffraction* and related phenomena that arise when the wavelength is *not* much shorter than all other length scales. We develop a general formalism for wave propagation based on the Helmholtz-Kirchhoff integral, which embodies Huygen's principle (the wave at a point is a sum over wavelets emanating from all locations on a surrounding sphere), which in turn is closely related to Feynman path integrals in quantum mechanics (sums over all possible histories). We use this formalism to explore diffraction patterns (Fresnel and Fraunhofer). Specializing to paraxial waves, we deduce a propagator for the waves, called the "point-spread function," that is closely related to quantum mechanical propagators. Among our applications of this propagator are diffraction near caustics; optical image processing; Gaussian beams of light and their manipulation; optical (Fabry-Perot) cavities for trapping and manipulating Gaussian beams; and interferometric gravitational-wave detectors, such as LIGO.

In Chap. 9, we focus on the superposition, merging, and *interference* of small numbers of beams of light or other waves and how the interference behaves, depending on the waves' degree of *coherence*. The formalism that we develop for quantifying coherence is a variant of the theory of random processes, studied in Volume 1 of MCP. Among our applications are various types of interferometers; spectrum-measuring devices based on interferometry (rather than on the usual diffraction gratings); radio telescope arrays, such as the Event Horizon Telescope; stabilization of laser frequencies by locking to an optical cavity; optical frequency combs and their practical applications; the twinkling of starlight, its influence on astronomical seeing, and adaptive optics to sharpen astronomical images; and a revisiting, in greater depth, of Gaussian beams, their manipulation, and LIGO.

Chapter 10 is devoted to nonlinear optics. Nonlinearities abound in the world around us (e.g., in avalanches, breaking ocean waves, and modern optical devices), but all too often they are ignored in physics curricula. In this chapter, we embrace nonlinearities and explore in depth nonlinear wave-wave mixing (the generation of new outgoing waves by nonlinear interactions of incoming waves). Such wave-wave mixing is the foundation for holography—a mature technique for recording and reconstructing 3-dimensional optical images. We explain holography and describe some of its modern variants and applications—holographic lenses, supermarket scanners, CDs, DVDs, BDs, and others. The technological foundation for wave-wave mixing is nonlinear media (e.g., nonlinear crystals). The properties of these media (most importantly, their index of refraction) change nonlinearly when two or more waves pass through them, and those changing properties trigger the incoming waves to generate new outgoing waves. In Chap. 10, we develop in detail the theory of this wave-wave mixing and use it to describe and analyze, for example, the (unwanted) nonlinear distortion of images as they propagate along optical fibers; phase-conjugate mirrors and their use to remove distortions from optical images; frequency-doubling crystals in laser pointers, which convert infrared light to green; optical parametric amplifiers; squeezers that amplify one quadrature of a light wave and attenuate the other quadrature; and the use of squeezers to manipulate the quantum electrodynamical vacuum and thereby control quantum noise.

Volume 4: Plasma Physics

Volume 5: Relativity and Cosmology

PART III

OPTICS

Prior to the twentieth century's quantum mechanics and opening of the electromagnetic spectrum observationally, the study of optics was concerned solely with visible light.

Reflection and refraction of light were first described by the Greeks and Arabs and further studied by such medieval scholastics as Roger Bacon (thirteenth century), who explained the rainbow and used refraction in the design of crude magnifying lenses and spectacles. However, it was not until the seventeenth century that there arose a strong commercial interest in manipulating light, particularly via the telescope and compound microscope, improved and famously used by Galileo and Newton.

The discovery of Snell's law in 1621 and observations of diffractive phenomena by Grimaldi in 1665 stimulated serious speculation about the physical nature of light. The wave and corpuscular theories were propounded by Huygens in 1678 and Newton in 1704, respectively. The corpuscular theory initially held sway, for 100 years. However, observational studies of interference by Young in 1803 and the derivation of a wave equation for electromagnetic disturbances by Maxwell in 1865 then seemed to settle the matter in favor of the undulatory theory, only for the debate to be resurrected in 1887 with the discovery of the photoelectric effect by Hertz. After quantum mechanics was developed in the 1920s, the dispute was abandoned, the wave and particle descriptions of light became "complementary," and Hamilton's optics-inspired formulation of classical mechanics was modified to produce the Schrödinger equation.

Many physics students are all too familiar with this potted history and may consequently regard optics as an ancient precursor to modern physics, one that has been completely subsumed by quantum mechanics. Not so! Optics has developed dramatically and independently from quantum mechanics in recent decades and is now a major branch of classical physics. And it is no longer concerned primarily with light. The principles of optics are routinely applied to all types of wave propagation: for example, all parts of the electromagnetic spectrum, quantum mechanical waves (e.g., of electrons and neutrinos), waves in elastic solids (Part IV of this book), fluids (Part V), plasmas (Part VI), and the geometry of spacetime (Part VII). There is a commonality, for instance, to seismology, oceanography, and radio physics that allows ideas to be freely interchanged among these different disciplines. Even the study of visible light has seen major developments: the invention of the laser has led to the modern theory of coherence and has begotten the new field of nonlinear optics.

An even greater revolution has occurred in optical technology. From the credit card and white-light hologram to the laser scanner at a supermarket checkout, from laser printers to CDs, DVDs, and BDs, from radio telescopes capable of nanoradian angular resolution to Fabry-Perot systems that detect displacements smaller than the size of an elementary particle, we are surrounded by sophisticated optical devices in our everyday and scientific lives. Many of these devices turn out to be clever and direct applications of the fundamental optical principles that we discuss in this part of the book.

Our treatment of optics in this part differs from that found in traditional texts, in that we assume familiarity with basic classical mechanics and quantum mechanics and, consequently, fluency in the language of Fourier transforms. This inversion of the historical development reflects contemporary priorities and allows us to emphasize those aspects of the subject that involve fresh concepts and modern applications.

In Chap. 7, we discuss optical (wave-propagation) phenomena in the geometric optics approximation. This approximation is accurate when the wavelength and the wave period are short compared with the lengthscales and timescales on which the wave amplitude and the waves' environment vary. We show how a wave equation can be solved approximately, with optical rays becoming the classical trajectories of quantum particles (photons, phonons, plasmons, and gravitons) and the wave field propagating along these trajectories. We also show how, in general, these trajectories develop singularities or caustics where the geometric optics approximation breaks down, and we must revert to the wave description.

In Chap. 8, we develop the theory of diffraction that arises when the geometric optics approximation fails, and the waves' energy spreads in a non-particle-like way. We analyze diffraction in two limiting regimes, called "Fresnel" and "Fraunhofer" (after the physicists who discovered them), in which the wavefronts are approximately planar or spherical, respectively. As we are working with a linear theory of wave propagation, we make heavy use of Fourier methods and show how elementary applications of Fourier transforms can be used to design powerful optics instruments.

Most elementary diffractive phenomena involve the superposition of an infinite number of waves. However, in many optical applications, only a small number of waves from a common source are combined. This is known as interference and is the subject of Chap. 9. In this chapter, we also introduce the notion of coherence, which is a quantitative measure of the distributions of the combining waves and their capacity to interfere constructively.

The final chapter on optics, Chap. 10, is concerned with nonlinear phenomena that arise when waves, propagating through a medium, become sufficiently strong to couple to one another. These nonlinear phenomena can occur for all types of waves (we meet them for fluid waves in Sec. 16.3 and plasma waves in Chap. 23). For light (the focus of Chap. 10), nonlinearities have become especially important in recent years; the nonlinear effects that arise when laser light is shone through certain crystals are having a strong impact on technology and on fundamental scientific research. We explore several examples.

CHAPTER SEVEN

Geometric Optics

Solar rays parallel to OB and passing through this solid are refracted at the hyperbolic surface, and the refracted rays converge at A.

IBN SAHL (984)

7.1 Overview

Geometric optics, the study of "rays," is the oldest approach to optics. It is an accurate description of wave propagation when the wavelengths and periods of the waves are far smaller than the lengthscales and timescales on which the wave amplitude and the medium supporting the waves vary.

After reviewing wave propagation in a homogeneous medium (Sec. 7.2), we begin our study of geometric optics in Sec. 7.3. There we derive the geometric-optics propagation equations with the aid of the eikonal approximation, and we elucidate the connection to Hamilton-Jacobi theory (which we assume the reader has already encountered). This connection is made more explicit by demonstrating that a classical, geometric-optics wave can be interpreted as a flux of quanta. In Sec. 7.4, we specialize the geometric-optics formalism to any situation where a bundle of nearly parallel rays is being guided and manipulated by some sort of apparatus. This is the paraxial approximation, and we illustrate it with a magnetically focused beam of charged particles and show how matrix methods can be used to describe the particle (i.e., ray) trajectories. In Sec. 7.5, we explore how imperfect optics can produce multiple images of a distant source, and that as one moves from one location to another, the images appear and disappear in pairs. Locations where this happens are called "caustics" and are governed by catastrophe theory, a topic we explore briefly. In Sec. 7.6, we describe gravitational lenses, remarkable astronomical phenomena that illustrate the formation of multiple images and caustics. Finally, in Sec. 7.7, we turn from scalar waves to the vector waves of electromagnetic radiation. We deduce the geometric-optics propagation law for the waves' polarization vector and explore the classical version of a phenomenon called geometric phase.

- This chapter does not depend substantially on any previous chapter, but it does assume familiarity with classical mechanics, quantum mechanics, and classical electromagnetism.
- Secs. 7.1–7.4 are foundations for the remaining optics chapters, 8, 9, and 10.
- The discussion of caustics in Sec. 7.5 is a foundation for Sec. 8.6 on diffraction at a caustic.
- Secs. 7.2 and 7.3 (monochromatic plane waves and wave packets in a homogeneous, time-independent medium; the dispersion relation; and the geometric-optics equations) are used extensively in subsequent parts of this book, including
  - Chap. 12 for elastodynamic waves,
  - Chap. 16 for waves in fluids,
  - Sec. 19.7 and Chaps. 21–23 for waves in plasmas, and
  - Chap. 27 for gravitational waves.
  - Sec. 28.6.2 for weak gravitational lensing.

7.2

7.2 Waves in a Homogeneous Medium

7.2.1

7.2.1 Monochromatic Plane Waves; Dispersion Relation

Consider a monochromatic plane wave propagating through a homogeneous medium. Independently of the physical nature of the wave, it can be described mathematically by

$$\psi = A e^{i(\mathbf{k} \cdot \mathbf{x} - \omega t)} \equiv A e^{i\varphi}, \tag{7.1}$$

plane wave: complex amplitude, phase, angular frequency, wave vector, wavelength, and propagation direction

where ψ is any oscillatory physical quantity associated with the wave, for example, the y component of the magnetic field associated with an electromagnetic wave. If, as is usually the case, the physical quantity is real (not complex), then we must take the real part of Eq. (7.1). In Eq. (7.1), A is the wave's *complex amplitude*; $\varphi = \mathbf{k} \cdot \mathbf{x} - \omega t$ is the wave's *phase*; t and $\mathbf{x}$ are time and location in space; $\omega = 2\pi f$ is the wave's *angular frequency*; and $\mathbf{k}$ is its *wave vector* (with $k \equiv |\mathbf{k}|$ its *wave number*, $\lambda = 2\pi/k$ its *wavelength*, $\lambda = \lambda/(2\pi)$ its *reduced wavelength*, and $\hat{\mathbf{k}} \equiv \mathbf{k}/k$ its *propagation direction*). Surfaces of constant phase φ are orthogonal to the propagation direction $\hat{\mathbf{k}}$ and move in the $\hat{\mathbf{k}}$ direction with the *phase velocity*

phase velocity

$$\mathbf{V}_{\text{ph}} \equiv \left(\frac{\partial \mathbf{x}}{\partial t} \right)_\varphi = \frac{\omega}{k} \hat{\mathbf{k}} \tag{7.2}$$

(cf. Fig. 7.1). The frequency ω is determined by the wave vector $\mathbf{k}$ in a manner that depends on the wave's physical nature; the functional relationship

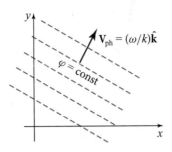

FIGURE 7.1 A monochromatic plane wave in a homogeneous medium.

$$\boxed{\omega = \Omega(\mathbf{k})} \tag{7.3}$$

is called the wave's *dispersion relation,* because (as we shall see in Ex. 7.2) it governs the dispersion (spreading) of a wave packet that is constructed by superposing plane waves.

Some examples of plane waves that we study in this book are:

1. Electromagnetic waves propagating through an isotropic dielectric medium with index of refraction $\mathfrak{n}$ [Eq. 10.20)], for which ψ could be any Cartesian component of the electric or magnetic field or vector potential and the dispersion relation is

$$\omega = \Omega(\mathbf{k}) = Ck \equiv C|\mathbf{k}|, \tag{7.4}$$

with $C = c/\mathfrak{n}$ the phase speed and c the speed of light in vacuum.

2. Sound waves propagating through a solid (Sec. 12.2.3) or fluid (liquid or vapor; Secs. 7.3.1 and 16.5), for which ψ could be the pressure or density perturbation produced by the sound wave (or it could be a potential whose gradient is the velocity perturbation), and the dispersion relation is the same as for electromagnetic waves, Eq. (7.4), but with C now the sound speed.

3. Waves on the surface of a deep body of water (depth $\gg \lambda$; Sec. 16.2.1), for which ψ could be the height of the water above equilibrium, and the dispersion relation is [Eq. (16.9)]:

$$\omega = \Omega(\mathbf{k}) = \sqrt{gk} = \sqrt{g|\mathbf{k}|}, \tag{7.5}$$

with g the acceleration of gravity.

4. Flexural waves on a stiff beam or rod (Sec. 12.3.4), for which ψ could be the transverse displacement of the beam from equilibrium, and the dispersion relation is

$$\omega = \Omega(\mathbf{k}) = \sqrt{\frac{D}{\Lambda}k^2} = \sqrt{\frac{D}{\Lambda}\mathbf{k} \cdot \mathbf{k}}, \tag{7.6}$$

with Λ the rod's mass per unit length and D its "flexural rigidity" [Eq. (12.33)].

5. Alfvén waves in a magnetized, nonrelativistic plasma (bending waves of the plasma-laden magnetic field lines; Sec. 19.7.2), for which ψ could be the transverse displacement of the field and plasma, and the dispersion relation is [Eq. (19.75)]

$$\omega = \Omega(\mathbf{k}) = \mathbf{a} \cdot \mathbf{k}, \tag{7.7}$$

with $\mathbf{a} = \mathbf{B}/\sqrt{\mu_o \rho}$, $[= \mathbf{B}/\sqrt{4\pi\rho}]^1$ the Alfvén speed, $\mathbf{B}$ the (homogeneous) magnetic field, μ_o the magnetic permittivity of the vacuum, and ρ the plasma mass density.

6. Gravitational waves propagating across the universe, for which ψ can be a component of the waves' metric perturbation which describes the waves' stretching and squeezing of space; these waves propagate nondispersively at the speed of light, so their dispersion relation is Eq. (7.4) with C replaced by the vacuum light speed c.

In general, one can derive the dispersion relation $\omega = \Omega(\mathbf{k})$ by inserting the plane-wave ansatz (7.1) into the dynamical equations that govern one's physical system [e.g., Maxwell's equations, the equations of elastodynamics (Chap. 12), or the equations for a magnetized plasma (Part VI)]. We shall do so time and again in this book.

7.2.2 Wave Packets

Waves in the real world are not precisely monochromatic and planar. Instead, they occupy wave packets that are somewhat localized in space and time. Such wave packets can be constructed as superpositions of plane waves:

$$\psi(\mathbf{x}, t) = \int A(\mathbf{k}) e^{i\alpha(\mathbf{k})} e^{i(\mathbf{k} \cdot \mathbf{x} - \omega t)} \frac{d^3k}{(2\pi)^3}, \tag{7.8a}$$

where $A(\mathbf{k})$ is concentrated around some $\mathbf{k} = \mathbf{k}_o$.

Here A and α (both real) are the modulus and phase of the complex amplitude $Ae^{i\alpha}$, and the integration element is $d^3k \equiv d\mathcal{V}_k \equiv dk_x dk_y dk_z$ in terms of components of $\mathbf{k}$ along Cartesian axes x, y, and z. In the integral (7.8a), the contributions from adjacent $\mathbf{k}$s will tend to cancel each other except in that region of space and time where the oscillatory phase factor changes little with changing $\mathbf{k}$ (when $\mathbf{k}$ is near $\mathbf{k}_o$). This is the spacetime region in which the wave packet is concentrated, and its center is where $\nabla_\mathbf{k}$(phase factor) $= 0$:

$$\left(\frac{\partial\alpha}{\partial k_j} + \frac{\partial}{\partial k_j}(\mathbf{k} \cdot \mathbf{x} - \omega t) \right)_{\mathbf{k}=\mathbf{k}_o} = 0. \tag{7.8b}$$

---

1. Gaussian unit equivalents will be given with square brackets.

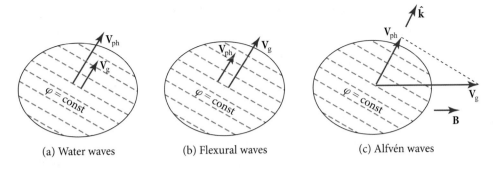

(a) Water waves   (b) Flexural waves   (c) Alfvén waves

FIGURE 7.2 (a) A wave packet of waves on a deep body of water. The packet is localized in the spatial region bounded by the ellipse. The packet's (ellipse's) center moves with the group velocity $\mathbf{V}_g$. The ellipse expands slowly due to wave-packet dispersion (spreading; Ex. 7.2). The surfaces of constant phase (the wave's oscillations) move twice as fast as the ellipse and in the same direction, $\mathbf{V}_{ph} = 2\mathbf{V}_g$ [Eq. (7.11)]. This means that the wave's oscillations arise at the back of the packet and move forward through the packet, disappearing at the front. The wavelength of these oscillations is $\lambda = 2\pi/k_o$, where $k_o = |\mathbf{k}_o|$ is the wave number about which the wave packet is concentrated [Eq. (7.8a) and associated discussion]. (b) A flexural wave packet on a beam, for which $\mathbf{V}_{ph} = \frac{1}{2}\mathbf{V}_g$ [Eq. (7.12)], so the wave's oscillations arise at the packet's front and, traveling more slowly than the packet, disappear at its back. (c) An Alfvén wave packet. Its center moves with a group velocity $\mathbf{V}_g$ that points along the direction of the background magnetic field [Eq. (7.13)], and its surfaces of constant phase (the wave's oscillations) move with a phase velocity $\mathbf{V}_{ph}$ that can be in any direction $\hat{\mathbf{k}}$. The phase speed is the projection of the group velocity onto the phase propagation direction, $|\mathbf{V}_{ph}| = \mathbf{V}_g \cdot \hat{\mathbf{k}}$ [Eq. (7.13)], which implies that the wave's oscillations remain fixed inside the packet as the packet moves; their pattern inside the ellipse does not change. (An even more striking example is provided by the Rossby wave, discussed in Sec. 16.4, in which the group velocity is equal and oppositely directed to the phase velocity.)

Evaluating the derivative with the aid of the wave's dispersion relation $\omega = \Omega(\mathbf{k})$, we obtain for the location of the wave packet's center

$$x_j - \left(\frac{\partial\Omega}{\partial k_j}\right)_{\mathbf{k}=\mathbf{k}_o} t = -\left(\frac{\partial\alpha}{\partial k_j}\right)_{\mathbf{k}=\mathbf{k}_o} = \text{const.} \tag{7.8c}$$

This tells us that the *wave packet* moves with the *group velocity*

$$\boxed{\mathbf{V}_g = \nabla_{\mathbf{k}}\Omega, \quad \text{i.e.,} \quad V_{gj} = \left(\frac{\partial\Omega}{\partial k_j}\right)_{\mathbf{k}=\mathbf{k}_o}.} \tag{7.9}$$

group velocity

When, as for electromagnetic waves in a dielectric medium or sound waves in a solid or fluid, the dispersion relation has the simple form of Eq. (7.4), $\omega = \Omega(\mathbf{k}) = Ck$ with $k \equiv |\mathbf{k}|$, then the group and phase velocities are the same,

$$\mathbf{V}_g = \mathbf{V}_{ph} = C\hat{\mathbf{k}}, \tag{7.10}$$

and the waves are said to be *dispersionless*. If the dispersion relation has any other form, then the group and phase velocities are different, and the wave is said to exhibit *dispersion;* cf. Ex. 7.2. Examples are (see Fig. 7.2 and the list in Sec. 7.2.1, from which our numbering is taken):

3. Waves on a deep body of water [dispersion relation (7.5); Fig. 7.2a], for which

$$\mathbf{V}_g = \frac{1}{2}\mathbf{V}_{ph} = \frac{1}{2}\sqrt{\frac{g}{k}}\,\hat{\mathbf{k}}; \tag{7.11}$$

4. Flexural waves on a stiff beam or rod [dispersion relation (7.6); Fig. 7.2b], for which

$$\mathbf{V}_g = 2\mathbf{V}_{ph} = 2\sqrt{\frac{D}{\Lambda}}\,k\hat{\mathbf{k}}; \tag{7.12}$$

5. Alfvén waves in a magnetized and nonrelativistic plasma [dispersion relation (7.7); Fig. 7.2c], for which

$$\mathbf{V}_g = \mathbf{a}, \qquad \mathbf{V}_{ph} = (\mathbf{a}\cdot\hat{\mathbf{k}})\hat{\mathbf{k}}. \tag{7.13}$$

Notice that, depending on the dispersion relation, the group speed $|\mathbf{V}_g|$ can be less than or greater than the phase speed, and if the homogeneous medium is anisotropic (e.g., for a magnetized plasma), the group velocity can point in a different direction than the phase velocity.

Physically, it should be obvious that the energy contained in a wave packet must remain always with the packet and cannot move into the region outside the packet where the wave amplitude vanishes. Correspondingly, the wave packet's energy must propagate with the group velocity $\mathbf{V}_g$ and not with the phase velocity $\mathbf{V}_{ph}$. When one examines the wave packet from a quantum mechanical viewpoint, its quanta must move with the group velocity $\mathbf{V}_g$. Since we have required that the wave packet have its wave vectors concentrated around $\mathbf{k}_o$, the energy and momentum of each of the packet's quanta are given by the standard quantum mechanical relations:

$$\boxed{\mathcal{E} = \hbar\Omega(\mathbf{k}_o), \quad \text{and} \quad \mathbf{p} = \hbar\mathbf{k}_o.} \tag{7.14}$$

EXERCISES

Exercise 7.1 *Practice: Group and Phase Velocities*
Derive the group and phase velocities (7.10)–(7.13) from the dispersion relations (7.4)–(7.7).

Exercise 7.2 **Example: Gaussian Wave Packet and Its Dispersion*
Consider a 1-dimensional wave packet, $\psi(x, t) = \int A(k)e^{i\alpha(k)}e^{i(kx-\omega t)}dk/(2\pi)$, with dispersion relation $\omega = \Omega(k)$. For concreteness, let $A(k)$ be a narrow Gaussian peaked around k_o: $A \propto \exp[-\kappa^2/(2(\Delta k)^2)]$, where $\kappa = k - k_o$.

(a) Expand α as $\alpha(k) = \alpha_o - x_o\kappa$ with x_o a constant, and assume for simplicity that higher order terms are negligible. Similarly, expand $\omega \equiv \Omega(k)$ to quadratic order,

and explain why the coefficients are related to the group velocity V_g at $k = k_o$ by
$\Omega = \omega_o + V_g \kappa + (dV_g/dk)\kappa^2/2$.

(b) Show that the wave packet is given by

$$\psi \propto \exp[i(\alpha_o + k_o x - \omega_o t)] \int_{-\infty}^{+\infty} \exp[i\kappa(x - x_o - V_g t)] \qquad (7.15a)$$

$$\times \exp\left[-\frac{\kappa^2}{2}\left(\frac{1}{(\Delta k)^2} + i\frac{dV_g}{dk}t\right)\right] d\kappa.$$

The term in front of the integral describes the phase evolution of the waves inside the packet; cf. Fig. 7.2.

(c) Evaluate the integral analytically (with the help of a computer, if you wish). From your answer, show that the modulus of ψ satisfies

$$\boxed{|\psi| \propto \frac{1}{L^{1/2}} \exp\left[-\frac{(x - x_o - V_g t)^2}{2L^2}\right], \quad \text{where } L = \frac{1}{\Delta k}\sqrt{1 + \left(\frac{dV_g}{dk}(\Delta k)^2 t\right)^2}}$$

$$(7.15b)$$

is the packet's half-width.

(d) Discuss the relationship of this result at time $t = 0$ to the uncertainty principle for the localization of the packet's quanta.

(e) Equation (7.15b) shows that the wave packet spreads (i.e., disperses) due to its containing a range of group velocities [Eq. (7.11)]. How long does it take for the packet to enlarge by a factor 2? For what range of initial half-widths can a water wave on the ocean spread by less than a factor 2 while traveling from Hawaii to California?

7.3 Waves in an Inhomogeneous, Time-Varying Medium: The Eikonal Approximation and Geometric Optics

7.3

Suppose that the medium in which the waves propagate is spatially inhomogeneous and varies with time. If the lengthscale $\mathcal{L}$ and timescale $\mathcal{T}$ for substantial variations are long compared to the waves' reduced wavelength and period,

$$\mathcal{L} \gg \lambdabar = 1/k, \qquad \mathcal{T} \gg 1/\omega, \qquad (7.16)$$

then the waves can be regarded locally as planar and monochromatic. The medium's inhomogeneities and time variations may produce variations in the wave vector $\mathbf{k}$ and frequency ω, but those variations should be substantial only on scales $\gtrsim \mathcal{L} \gg 1/k$ and $\gtrsim \mathcal{T} \gg 1/\omega$. This intuitively obvious fact can be proved rigorously using a two-lengthscale expansion (i.e., an expansion of the wave equation in powers of $\lambdabar/\mathcal{L} = 1/k\mathcal{L}$ and $1/\omega\mathcal{T}$). Such an expansion, in this context of wave propagation, is called

two-lengthscale expansion

the *geometric-optics approximation* or the *eikonal approximation* (after the Greek word $\epsilon\iota\kappa\omega\nu$, meaning image). When the waves are those of elementary quantum mechanics, it is called the *WKB approximation*.[2] The eikonal approximation converts the laws of wave propagation into a remarkably simple form, in which the waves' amplitude is transported along trajectories in spacetime called *rays*. In the language of quantum mechanics, these rays are the world lines of the wave's quanta (photons for light, phonons for sound, plasmons for Alfvén waves, and gravitons for gravitational waves), and the law by which the wave amplitude is transported along the rays is one that conserves quanta. These ray-based propagation laws are called the laws of *geometric optics*.

In this section we develop and study the eikonal approximation and its resulting laws of geometric optics. We begin in Sec. 7.3.1 with a full development of the eikonal approximation and its geometric-optics consequences for a prototypical dispersion-free wave equation that represents, for example, sound waves in a weakly inhomogeneous fluid. In Sec. 7.3.3, we extend our analysis to cover all other types of waves. In Sec. 7.3.4 and a number of exercises we explore examples of geometric-optics waves, and in Sec. 7.3.5 we discuss conditions under which the eikonal approximation breaks down and some non-geometric-optics phenomena that result from the breakdown. Finally, in Sec. 7.3.6 we return to nondispersive light and sound waves, deduce Fermat's principle, and explore some of its consequences.

7.3.1 Geometric Optics for a Prototypical Wave Equation

Our prototypical wave equation is

$$\frac{\partial}{\partial t}\left(W\frac{\partial\psi}{\partial t}\right) - \nabla\cdot(WC^2\nabla\psi) = 0. \tag{7.17}$$

Here $\psi(\mathbf{x}, t)$ is the quantity that oscillates (the *wave field*), $C(\mathbf{x}, t)$ will turn out to be the wave's slowly varying *propagation speed*, and $W(\mathbf{x}, t)$ is a slowly varying *weighting function* that depends on the properties of the medium through which the wave propagates. As we shall see, W has no influence on the wave's dispersion relation or on its geometric-optics rays, but it does influence the law of transport for the waves' amplitude.

The wave equation (7.17) describes sound waves propagating through a static, isentropic, inhomogeneous fluid (Ex. 16.13), in which case ψ is the wave's pressure perturbation δP, $C(\mathbf{x}) = \sqrt{(\partial P/\partial\rho)_s}$ is the adiabatic sound speed, and the weighting function is $W(\mathbf{x}) = 1/(\rho C^2)$, with ρ the fluid's unperturbed density. This wave equation also describes waves on the surface of a lake or pond or the ocean, in the limit that the slowly varying depth of the undisturbed water $h_o(\mathbf{x})$ is small compared

---

2. Sometimes called "JWKB," adding Jeffreys to the attribution, though Carlini, Liouville, and Green used it a century earlier.

to the wavelength (shallow-water waves; e.g., tsunamis); see Ex. 16.3. In this case ψ is the perturbation of the water's depth, $W = 1$, and $C = \sqrt{gh_o}$ with g the acceleration of gravity. In both cases—sound waves in a fluid and shallow-water waves—if we turn on a slow time dependence in the unperturbed fluid, then additional terms enter the wave equation (7.17). For pedagogical simplicity we leave those terms out, but in the analysis below we do allow W and C to be slowly varying in time, as well as in space: $W = W(\mathbf{x}, t)$ and $C = C(\mathbf{x}, t)$.

Associated with the wave equation (7.17) are an energy density $U(\mathbf{x}, t)$ and energy flux $\mathbf{F}(\mathbf{x}, t)$ given by

$$U = W \left[\frac{1}{2}\left(\frac{\partial \psi}{\partial t}\right)^2 + \frac{1}{2}C^2(\nabla\psi)^2 \right], \qquad \mathbf{F} = -WC^2 \frac{\partial \psi}{\partial t}\nabla\psi; \qquad (7.18)$$

energy density and flux

see Ex. 7.4. It is straightforward to verify that, if C and W are independent of time t, then the scalar wave equation (7.17) guarantees that the U and $\mathbf{F}$ of Eq. (7.18) satisfy the law of energy conservation:

$$\frac{\partial U}{\partial t} + \nabla \cdot \mathbf{F} = 0; \qquad (7.19)$$

cf. Ex. 7.4.[3]

We now specialize to a weakly inhomogeneous and slowly time-varying fluid and to nearly plane waves, and we seek a solution of the wave equation (7.17) that locally has approximately the plane-wave form $\psi \simeq A e^{i\mathbf{k}\cdot\mathbf{x}-\omega t}$. Motivated by this plane-wave form, (i) we express the waves in the eikonal approximation as the product of a real amplitude $A(\mathbf{x}, t)$ that varies slowly on the length- and timescales $\mathcal{L}$ and $\mathcal{T}$, and the exponential of a complex phase $\varphi(\mathbf{x}, t)$ that varies rapidly on the timescale $1/\omega$ and lengthscale λ:

eikonal approximated wave: amplitude, phase, wave vector, and angular frequency

$$\psi(\mathbf{x}, t) = A(\mathbf{x}, t)e^{i\varphi(\mathbf{x},t)}; \qquad (7.20)$$

and (ii) we define the wave vector (field) and angular frequency (field) by

$$\mathbf{k}(\mathbf{x}, t) \equiv \nabla\varphi, \qquad \omega(\mathbf{x}, t) \equiv -\partial\varphi/\partial t. \qquad (7.21)$$

In addition to our two-lengthscale requirement, $\mathcal{L} \gg 1/k$ and $\mathcal{T} \gg 1/\omega$, we also require that A, $\mathbf{k}$, and ω vary slowly (i.e., vary on lengthscales $\mathcal{R}$ and timescales $\mathcal{T}'$ long compared to $\lambda = 1/k$ and $1/\omega$).[4] This requirement guarantees that the waves are locally planar, $\varphi \simeq \mathbf{k} \cdot x - \omega t + \text{constant}$.

---

3. Alternatively, one can observe that a stationary medium will not perform work.
4. Note that these variations can arise both (i) from the influence of the medium's inhomogeneity (which puts limits $\mathcal{R} \lesssim \mathcal{L}$ and $\mathcal{T}' \lesssim \mathcal{T}$ on the wave's variations) and (ii) from the chosen form of the wave. For example, the wave might be traveling outward from a source and so have nearly spherical phase fronts with radii of curvature $r \simeq$ (distance from source); then $\mathcal{R} = \min(r, \mathcal{L})$.

BOX 7.2.   BOOKKEEPING PARAMETER IN TWO-LENGTHSCALE EXPANSIONS

When developing a two-lengthscale expansion, it is sometimes helpful to introduce a bookkeeping parameter σ and rewrite the ansatz (7.20) in a fleshed-out form:

$$\psi = (A + \sigma B + \ldots)e^{i\varphi/\sigma}. \tag{1}$$

The numerical value of σ is unity, so it can be dropped when the analysis is finished. We use σ to tell us how various terms scale when λ is reduced at fixed $\mathcal{L}$ and $\mathcal{R}$. The amplitude A has no attached σ and so scales as λ^0, B is multiplied by σ and so scales proportional to λ, and φ is multiplied by σ^{-1} and so scales as λ^{-1}. When one uses these factors of σ in the evaluation of the wave equation, the first term on the right-hand side of Eq. (7.22) gets multiplied by σ^{-2}, the second term by σ^{-1}, and the omitted terms by σ^0. These factors of σ help us to quickly group together all terms that scale in a similar manner and to identify which of the groupings is leading order, and which subleading, in the two-lengthscale expansion. In Eq. (7.22) the omitted σ^0 terms are the first ones in which B appears; they produce a propagation law for B, which can be regarded as a post-geometric-optics correction.

Occasionally the wave equation itself will contain terms that scale with λ differently from one another (e.g., Ex. 7.9). One should always look out for this possibility.

We now insert the eikonal-approximated wave field (7.20) into the wave equation (7.17), perform the differentiations with the aid of Eqs. (7.21), and collect terms in a manner dictated by a two-lengthscale expansion (see Box 7.2):

$$0 = \frac{\partial}{\partial t}\left(W\frac{\partial\psi}{\partial t}\right) - \nabla\cdot(WC^2\nabla\psi) \tag{7.22}$$

$$= \left(-\omega^2 + C^2k^2\right)W\psi$$

$$+ i\left[-2\left(\omega\frac{\partial A}{\partial t} + C^2k_jA_{,j}\right)W - \frac{\partial(W\omega)}{\partial t}A - (WC^2k_j)_{,j}A\right]e^{i\varphi} + \cdots.$$

The first term on the right-hand side, $(-\omega^2 + C^2k^2)W\psi$, scales as λ^{-2} when we make the reduced wavelength λ shorter and shorter while holding the macroscopic lengthscales $\mathcal{L}$ and $\mathcal{R}$ fixed; the second term (in square brackets) scales as λ^{-1}; and the omitted terms scale as λ^0. This is what we mean by "collecting terms in a manner dictated by a two-lengthscale expansion." Because of their different scaling, the first, second, and omitted terms must vanish separately; they cannot possibly cancel one another.

The vanishing of the first term in the eikonal-approximated wave equation (7.22) implies that the waves' frequency field $\omega(\mathbf{x}, t) \equiv -\partial\varphi/\partial t$ and wave-vector field $\mathbf{k} \equiv \nabla\varphi$ satisfy the dispersionless dispersion relation,

$$\omega = \Omega(\mathbf{k}, \mathbf{x}, t) \equiv C(\mathbf{x}, t)k, \tag{7.23}$$

where (as throughout this chapter) $k \equiv |\mathbf{k}|$. Notice that, as promised, this dispersion relation is independent of the weighting function W in the wave equation. Notice further that this dispersion relation is identical to that for a precisely plane wave in a homogeneous medium, Eq. (7.4), except that the propagation speed C is now a slowly varying function of space and time. This will always be so.

One can always deduce the geometric-optics dispersion relation by (i) considering a precisely plane, monochromatic wave in a precisely homogeneous, time-independent medium and deducing $\omega = \Omega(\mathbf{k})$ in a functional form that involves the medium's properties (e.g., density) and then (ii) allowing the properties to be slowly varying functions of $\mathbf{x}$ and t. The resulting dispersion relation [e.g., Eq. (7.23)] then acquires its $\mathbf{x}$ and t dependence from the properties of the medium.

The vanishing of the second term in the eikonal-approximated wave equation (7.22) dictates that the wave's real amplitude A is transported with the group velocity $\mathbf{V}_g = C\hat{\mathbf{k}}$ in the following manner:

$$\frac{dA}{dt} \equiv \left(\frac{\partial}{\partial t} + \mathbf{V}_g \cdot \nabla\right) A = -\frac{1}{2W\omega}\left[\frac{\partial(W\omega)}{\partial t} + \nabla \cdot (WC^2\mathbf{k})\right] A. \tag{7.24}$$

This propagation law, by contrast with the dispersion relation, does depend on the weighting function W. We return to this propagation law shortly and shall understand more deeply its dependence on W, but first we must investigate in detail the directions along which A is transported.

The time derivative $d/dt = \partial/\partial t + \mathbf{V}_g \cdot \nabla$ appearing in the propagation law (7.24) is similar to the derivative with respect to proper time along a world line in special relativity, $d/d\tau = u^0\partial/\partial t + \mathbf{u} \cdot \nabla$ (with u^α the world line's 4-velocity). This analogy tells us that the waves' amplitude A is being propagated along some sort of world lines (trajectories). Those world lines (the waves' rays), in fact, are governed by Hamilton's equations of particle mechanics with the dispersion relation $\Omega(\mathbf{x}, t, \mathbf{k})$ playing the role of the hamiltonian and $\mathbf{k}$ playing the role of momentum:

$$\boxed{\frac{dx_j}{dt} = \left(\frac{\partial\Omega}{\partial k_j}\right)_{\mathbf{x},t} \equiv V_{g\,j},} \quad \boxed{\frac{dk_j}{dt} = -\left(\frac{\partial\Omega}{\partial x_j}\right)_{\mathbf{k},t},} \quad \boxed{\frac{d\omega}{dt} = \left(\frac{\partial\Omega}{\partial t}\right)_{\mathbf{x},\mathbf{k}}.} \tag{7.25}$$

The first of these Hamilton equations is just our definition of the group velocity, with which [according to Eq. (7.24)] the amplitude is transported. The second tells us how

dispersion relation

propagation law for amplitude

rays

Hamilton's equations for rays

the wave vector $\mathbf{k}$ changes along a ray, and together with our knowledge of $C(\mathbf{x}, t)$, it tells us how the group velocity $\mathbf{V}_g = C\hat{\mathbf{k}}$ for our dispersionless waves changes along a ray, and thence defines the ray itself. The third tells us how the waves' frequency changes along a ray.

To deduce the second and third of these Hamilton equations, we begin by inserting the definitions $\omega = -\partial\varphi/\partial t$ and $\mathbf{k} = \boldsymbol{\nabla}\varphi$ [Eqs. (7.21)] into the dispersion relation $\omega = \Omega(\mathbf{x}, t; \mathbf{k})$ for an arbitrary wave, thereby obtaining

$$\boxed{\frac{\partial\varphi}{\partial t} + \Omega(\mathbf{x}, t; \boldsymbol{\nabla}\varphi) = 0.} \tag{7.26a}$$

eikonal equation and Hamilton-Jacobi equation

This equation is known in optics as the *eikonal equation*. It is formally the same as the Hamilton-Jacobi equation of classical mechanics (see, e.g., Goldstein, Poole, and Safko, 2002), if we identify Ω with the hamiltonian and φ with Hamilton's principal function (cf. Ex. 7.9). This suggests that, to derive the second and third of Eqs. (7.25), we can follow the same procedure as is used to derive Hamilton's equations of motion. We take the gradient of Eq. (7.26a) to obtain

$$\frac{\partial^2\varphi}{\partial t \partial x_j} + \frac{\partial\Omega}{\partial k_l}\frac{\partial^2\varphi}{\partial x_l \partial x_j} + \frac{\partial\Omega}{\partial x_j} = 0, \tag{7.26b}$$

where the partial derivatives of Ω are with respect to its arguments $(\mathbf{x}, t; \mathbf{k})$; we then use $\partial\varphi/\partial x_j = k_j$ and $\partial\Omega/\partial k_l = V_{g\,l}$ to write Eq. (7.26b) as $dk_j/dt = -\partial\Omega/\partial x_j$. This is the second of Hamilton's equations (7.25), and it tells us how the wave vector changes along a ray. The third Hamilton equation, $d\omega/dt = \partial\Omega/\partial t$ [Eq. (7.25)], is obtained by taking the time derivative of the eikonal equation (7.26a).

Not only is the waves' amplitude A propagated along the rays, so also is their phase:

propagation equation for phase

$$\frac{d\varphi}{dt} = \frac{\partial\varphi}{\partial t} + \mathbf{V}_g \cdot \boldsymbol{\nabla}\varphi = -\omega + \mathbf{V}_g \cdot \mathbf{k}. \tag{7.27}$$

Since our dispersionless waves have $\omega = Ck$ and $\mathbf{V}_g = C\hat{\mathbf{k}}$, this vanishes. Therefore, for the special case of dispersionless waves (e.g., sound waves in a fluid and electromagnetic waves in an isotropic dielectric medium), the phase is constant along each ray:

$$\boxed{d\varphi/dt = 0.} \tag{7.28}$$

7.3.2

7.3.2 Connection of Geometric Optics to Quantum Theory

Although the waves $\psi = Ae^{i\varphi}$ are classical and our analysis is classical, their propagation laws in the eikonal approximation can be described most nicely in quantum mechanical language.[5] Quantum mechanics insists that, associated with any wave in

---

5. This is intimately related to the fact that quantum mechanics underlies classical mechanics; the classical world is an approximation to the quantum world, often a very good approximation.

the geometric-optics regime, there are real quanta: the wave's quantum mechanical particles. If the wave is electromagnetic, the quanta are photons; if it is gravitational, they are gravitons; if it is sound, they are phonons; if it is a plasma wave (e.g., Alfvén), they are plasmons. When we multiply the wave's $\mathbf{k}$ and ω by $\hbar$, we obtain the particles' momentum and energy:

quanta

$$\mathbf{p} = \hbar\mathbf{k}, \qquad \mathcal{E} = \hbar\omega. \tag{7.29}$$

momentum and energy of quanta

Although the originators of the nineteenth-century theory of classical waves were unaware of these quanta, once quantum mechanics had been formulated, the quanta became a powerful conceptual tool for thinking about classical waves.

In particular, we can regard the rays as the world lines of the quanta, and by multiplying the dispersion relation by $\hbar$, we can obtain the hamiltonian for the quanta's world lines:

$$H(\mathbf{x}, t; \mathbf{p}) = \hbar\Omega(\mathbf{x}, t; \mathbf{k} = \mathbf{p}/\hbar). \tag{7.30}$$

hamiltonian for quanta

Hamilton's equations (7.25) for the rays then immediately become Hamilton's equations for the quanta: $dx_j/dt = \partial H/\partial p_j$, $dp_j/dt = -\partial H/\partial x_j$, and $d\mathcal{E}/dt = \partial H/\partial t$.

Return now to the propagation law (7.24) for the waves' amplitude, and examine its consequences for the waves' energy. By inserting the ansatz $\psi = \Re(Ae^{i\varphi}) = A\cos(\varphi)$ into Eqs. (7.18) for the energy density U and energy flux $\mathbf{F}$ and averaging over a wavelength and wave period (so $\overline{\cos^2\varphi} = \overline{\sin^2\varphi} = 1/2$), we find that

$$U = \frac{1}{2}WC^2k^2A^2 = \frac{1}{2}W\omega^2A^2, \qquad \mathbf{F} = U(C\hat{\mathbf{k}}) = U\mathbf{V}_{\mathrm{g}}. \tag{7.31}$$

Inserting these into the expression $\partial U/\partial t + \nabla \cdot \mathbf{F}$ for the rate at which energy (per unit volume) fails to be conserved and using the propagation law (7.24) for A, we obtain

$$\frac{\partial U}{\partial t} + \nabla \cdot \mathbf{F} = U\frac{\partial \ln C}{\partial t}. \tag{7.32}$$

Thus, as the propagation speed C slowly changes at a fixed location in space due to a slow change in the medium's properties, the medium slowly pumps energy into the waves or removes it from them at a rate per unit volume of $U\partial \ln C/\partial t$.

This slow energy change can be understood more deeply using quantum concepts. The number density and number flux of quanta are

$$n = \frac{U}{\hbar\omega}, \qquad \mathbf{S} = \frac{\mathbf{F}}{\hbar\omega} = n\mathbf{V}_{\mathrm{g}}. \tag{7.33}$$

number density and flux for quanta

By combining these equations with the energy (non)conservation equation (7.32), we obtain

$$\frac{\partial n}{\partial t} + \nabla \cdot \mathbf{S} = n\left[\frac{\partial \ln C}{\partial t} - \frac{d \ln \omega}{dt}\right]. \tag{7.34}$$

The third Hamilton equation (7.25) tells us that

$$d\omega/dt = (\partial\Omega/\partial t)_{x,k} = [\partial(Ck)/\partial t]_{x,k} = k\partial C/\partial t,$$

whence $d \ln \omega/dt = \partial \ln C/\partial t$, which, when inserted into Eq. (7.34), implies that the quanta are conserved:

$$\boxed{\frac{\partial n}{\partial t} + \nabla \cdot \mathbf{S} = 0.}$$

(7.35a)

Since $\mathbf{S} = n\mathbf{V}_g$ and $d/dt = \partial/\partial t + \mathbf{V}_g \cdot \nabla$, we can rewrite this conservation law as a propagation law for the number density of quanta:

$$\boxed{\frac{dn}{dt} + n\nabla \cdot \mathbf{V}_g = 0.}$$

(7.35b)

The propagation law for the waves' amplitude, Eq. (7.24), can now be understood much more deeply: *The amplitude propagation law is nothing but the law of conservation of quanta in a slowly varying medium, rewritten in terms of the amplitude. This is true quite generally, for any kind of wave (Sec. 7.3.3); and the quickest route to the amplitude propagation law is often to express the wave's energy density U in terms of the amplitude and then invoke conservation of quanta*, Eq. (7.35b).

In Ex. 7.3 we show that the conservation law (7.35b) is equivalent to

$$\boxed{\frac{d(nC\mathcal{A})}{dt} = 0, \quad \text{i.e., } nC\mathcal{A} \text{ is a constant along each ray.}}$$

(7.35c)

Here $\mathcal{A}$ is the cross sectional area of a bundle of rays surrounding the ray along which the wave is propagating. Equivalently, by virtue of Eqs. (7.33) and (7.31) for the number density of quanta in terms of the wave amplitude A, we have

$$\frac{d}{dt} A\sqrt{CW\omega\mathcal{A}} = 0, \quad \text{i.e., } A\sqrt{CW\omega\mathcal{A}} \text{ is a constant along each ray.}$$

(7.35d)

In Eqs. (7.33) and (7.35), we have boxed those equations that are completely general (because they embody conservation of quanta) and have not boxed those that are specialized to our prototypical wave equation.

Exercise 7.3 ** *Derivation and Example: Amplitude Propagation for Dispersionless Waves Expressed as Constancy of Something along a Ray*

(a) In connection with Eq. (7.35b), explain why $\nabla \cdot \mathbf{V}_g = d \ln \mathcal{V}/dt$, where $\mathcal{V}$ is the tiny volume occupied by a collection of the wave's quanta.

(b) Choose for the collection of quanta those that occupy a cross sectional area $\mathcal{A}$ orthogonal to a chosen ray, and a longitudinal length Δs along the ray, so $\mathcal{V} = \mathcal{A}\Delta s$. Show that $d \ln \Delta s/dt = d \ln C/dt$ and correspondingly, $d \ln \mathcal{V}/dt = d \ln(C\mathcal{A})/dt$.

(c) Given part (b), show that the conservation law (7.35b) is equivalent to the constancy of $nC\mathcal{A}$ along a ray, Eq. (7.35c).

(d) From the results of part (c), derive the constancy of $A\sqrt{CW\omega\mathcal{A}}$ along a ray (where A is the wave's amplitude), Eq. (7.35d).

Exercise 7.4 **Example: Energy Density and Flux, and Adiabatic Invariant,*
for a Dispersionless Wave

(a) Show that the prototypical scalar wave equation (7.17) follows from the variational principle

$$\delta \int \mathcal{L}\,dt\,d^3x = 0, \qquad (7.36a)$$

where $\mathcal{L}$ is the lagrangian density

$$\mathcal{L} = W\left[\frac{1}{2}\left(\frac{\partial\psi}{\partial t}\right)^2 - \frac{1}{2}C^2\,(\nabla\psi)^2\right] \qquad (7.36b)$$

(not to be confused with the lengthscale $\mathcal{L}$ of inhomogeneities in the medium).

(b) For any scalar-field lagrangian density $\mathcal{L}(\psi, \partial\psi/\partial t, \nabla\psi, \mathbf{x}, t)$, the energy density and energy flux can be expressed in terms of the lagrangian, in Cartesian coordinates, as

$$U(\mathbf{x}, t) = \frac{\partial\psi}{\partial t}\frac{\partial\mathcal{L}}{\partial\psi/\partial t} - \mathcal{L}, \qquad F_j = \frac{\partial\psi}{\partial t}\frac{\partial\mathcal{L}}{\partial\psi/\partial x_j} \qquad (7.36c)$$

(Goldstein, Poole, and Safko, 2002, Sec. 13.3). Show, from the Euler-Lagrange equations for $\mathcal{L}$, that these expressions satisfy energy conservation, $\partial U/\partial t + \nabla\cdot\mathbf{F} = 0$, if $\mathcal{L}$ has no explicit time dependence [e.g., for the lagrangian (7.36b) if $C = C(\mathbf{x})$ and $W = W(\mathbf{x})$ do not depend on time t].

(c) Show that expression (7.36c) for the field's energy density U and its energy flux F_j agree with Eqs. (7.18).

(d) Now, regard the wave amplitude ψ as a generalized (field) coordinate. Use the lagrangian $L = \int \mathcal{L}\,d^3x$ to define a field momentum Π conjugate to this ψ, and then compute a *wave action*,

$$J \equiv \int_0^{2\pi/\omega} \int \Pi(\partial\psi/\partial t)\,d^3x\,dt, \qquad (7.36d)$$

which is the continuum analog of Eq. (7.43) in Sec. 7.3.6. The temporal integral is over one wave period. Show that this J is proportional to the wave energy divided by the frequency and thence to the number of quanta in the wave.

　　It is shown in standard texts on classical mechanics that, for approximately periodic oscillations, the particle action (7.43), with the integral limited to one period of oscillation of q, is an *adiabatic invariant*. By the extension of that proof to continuum physics, the wave action (7.36d) is also an adiabatic invariant. This means that the wave action and hence the number of quanta in the waves are conserved when the medium [in our case the index of refraction $n(\mathbf{x})$] changes very slowly in time—a result asserted in the text, and one that also follows from quantum mechanics. We study the particle version (7.43) of this adiabatic invariant in detail when we analyze charged-particle motion in a slowly varying magnetic field in Sec. 20.7.4.

Exercise 7.5 *Problem: Propagation of Sound Waves in a Wind*
Consider sound waves propagating in an atmosphere with a horizontal wind. Assume that the sound speed C, as measured in the air's local rest frame, is constant. Let the wind velocity $\mathbf{u} = u_x \mathbf{e}_x$ increase linearly with height z above the ground: $u_x = Sz$, where S is the constant shearing rate. Consider only rays in the x-z plane.

(a) Give an expression for the dispersion relation $\omega = \Omega(\mathbf{x}, t; \mathbf{k})$. [Hint: In the local rest frame of the air, Ω should have its standard sound-wave form.]

(b) Show that k_x is constant along a ray path, and then demonstrate that sound waves will not propagate when

$$\left| \frac{\omega}{k_x} - u_x(z) \right| < C. \tag{7.37}$$

(c) Consider sound rays generated on the ground that make an angle θ to the horizontal initially. Derive the equations describing the rays, and use them to sketch the rays, distinguishing values of θ both less than and greater than $\pi/2$. (You might like to perform this exercise numerically.)

7.3.3 Geometric Optics for a General Wave

With the simple case of nondispersive sound waves (Secs. 7.3.1 and 7.3.2) as our model, we now study an arbitrary kind of wave in a weakly inhomogeneous and slowly time varying medium (e.g., any of the examples in Sec. 7.2.1: light waves in a dielectric medium, deep water waves, flexural waves on a stiff beam, or Alfvén waves). Whatever the wave may be, we seek a solution to its wave equation using the eikonal approximation $\psi = Ae^{i\varphi}$ with slowly varying amplitude A and rapidly varying phase φ. Depending on the nature of the wave, ψ and A might be a scalar (e.g., sound waves), a vector (e.g., light waves), or a tensor (e.g., gravitational waves).

When we insert the ansatz $\psi = Ae^{i\varphi}$ into the wave equation and collect terms in the manner dictated by our two-lengthscale expansion [as in Eq. (7.22) and Box 7.2], the leading-order term will arise from letting every temporal or spatial derivative act on the $e^{i\varphi}$. This is precisely where the derivatives would operate in the case of a plane wave in a homogeneous medium, and here, as there, the result of each differentiation is $\partial e^{i\varphi}/\partial t = -i\omega e^{i\varphi}$ or $\partial e^{i\varphi}/\partial x_j = ik_j e^{i\varphi}$. Correspondingly, the leading-order terms in the wave equation here will be identical to those in the homogeneous plane wave case: they will be the dispersion relation multiplied by something times the wave,

$$[-\omega^2 + \Omega^2(\mathbf{x}, t; \mathbf{k})] \times (\text{something}) A e^{i\varphi} = 0, \qquad (7.38a)$$

dispersion relation for general wave

with the spatial and temporal dependence of Ω^2 entering through the medium's properties. This guarantees that (as we claimed in Sec. 7.3.1) the dispersion relation can be obtained by analyzing a plane, monochromatic wave in a homogeneous, time-independent medium and then letting the medium's properties, in the dispersion relation, vary slowly with $\mathbf{x}$ and t.

Each next-order ("subleading") term in the wave equation will entail just one of the wave operator's derivatives acting on a slowly varying quantity (A, a medium property, ω, or $\mathbf{k}$) and all the other derivatives acting on $e^{i\varphi}$. The subleading terms that interest us, for the moment, are those in which the one derivative acts on A, thereby propagating it. Therefore, the subleading terms can be deduced from the leading-order terms (7.38a) by replacing just one $i\omega A e^{i\varphi} = -A(e^{i\varphi})_{,t}$ by $-A_{,t}e^{i\varphi}$, and replacing just one $ik_j A e^{i\varphi} = A(e^{i\varphi})_{,j}$ by $A_{,j}e^{i\varphi}$ (where the subscript commas denote partial derivatives in Cartesian coordinates). A little thought then reveals that the equation for the vanishing of the subleading terms must take the form [deducible from the leading terms (7.38a)]:

$$-2i\omega\frac{\partial A}{\partial t} - 2i\Omega(\mathbf{k}, \mathbf{x}, t)\frac{\partial\Omega(\mathbf{k}, \mathbf{x}, t)}{\partial k_j}\frac{\partial A}{\partial x_j} = \text{terms proportional to } A. \quad (7.38b)$$

Using the dispersion relation $\omega = \Omega(\mathbf{x}, t; \mathbf{k})$ and the group velocity (first Hamilton equation) $\partial\Omega/\partial k_j = V_{g\,j}$, we bring this into the "propagate A along a ray" form:

$$\frac{dA}{dt} \equiv \frac{\partial A}{\partial t} + \mathbf{V}_{\mathrm{g}} \cdot \boldsymbol{\nabla} A = \text{terms proportional to } A. \qquad (7.38c)$$

Let us return to the leading-order terms (7.38a) in the wave equation [i.e., to the dispersion relation $\omega = \Omega(\mathbf{x}, t; \mathbf{k})$]. For our general wave, as for the prototypical dispersionless wave of the previous two sections, the argument embodied in Eqs. (7.26) shows that the rays are determined by Hamilton's equations (7.25),

$$\boxed{\frac{dx_j}{dt} = \left(\frac{\partial\Omega}{\partial k_j}\right)_{\mathbf{x},t} \equiv V_{g\,j}, \quad \frac{dk_j}{dt} = -\left(\frac{\partial\Omega}{\partial x_j}\right)_{\mathbf{k},t}, \quad \frac{d\omega}{dt} = \left(\frac{\partial\Omega}{\partial t}\right)_{\mathbf{x},\mathbf{k}},} \quad (7.39)$$

Hamilton's equations for general wave

but using the general wave's dispersion relation $\Omega(\mathbf{k}, \mathbf{x}, t)$ rather than $\Omega = C(\mathbf{x}, t)k$. These Hamilton equations include propagation laws for $\omega = -\partial\varphi/\partial t$ and $k_j = \partial\varphi/\partial x_j$, from which we can deduce the propagation law (7.27) for φ along the rays:

$$\boxed{\frac{d\varphi}{dt} = -\omega + \mathbf{V}_{\mathrm{g}} \cdot \mathbf{k}.} \qquad (7.40)$$

propagation law for phase of general wave

For waves with dispersion, by contrast with sound in a fluid and other waves that have $\Omega = Ck$, φ will not be constant along a ray.

For our general wave, as for dispersionless waves, the Hamilton equations for the rays can be reinterpreted as Hamilton's equations for the world lines of the waves' quanta [Eq. (7.30) and associated discussion]. And for our general wave, as for dispersionless waves, the medium's slow variations are incapable of creating or destroying wave quanta.[6] Correspondingly, if one knows the relationship between the waves' energy density U and their amplitude A, and thence the relationship between the waves' quantum number density $n = U/\hbar\omega$ and A, then *from the quantum conservation law* [boxed Eqs. (7.35)]

conservation of quanta and propagation of amplitude for general wave

$$\frac{\partial n}{\partial t} + \boldsymbol{\nabla} \cdot (n\mathbf{V}_g) = 0, \quad \frac{dn}{dt} + n\boldsymbol{\nabla} \cdot \mathbf{V}_g = 0, \quad \text{or} \quad \frac{d(nC\mathcal{A})}{dt} = 0, \quad (7.41)$$

one can deduce the propagation law for A—and the result must be the same propagation law as one obtains from the subleading terms in the eikonal approximation.

7.3.4 Examples of Geometric-Optics Wave Propagation

SPHERICAL SOUND WAVES

As a simple example of these geometric-optics propagation laws, consider a sound wave propagating radially outward through a homogeneous fluid from a spherical source (e.g., a radially oscillating ball; cf. Sec. 16.5.3). The dispersion relation is Eq. (7.4): $\Omega = Ck$. It is straightforward (Ex. 7.6) to integrate Hamilton's equations and learn that the rays have the simple form $\{r = Ct + \text{constant}, \theta = \text{constant}, \phi = \text{constant}, \mathbf{k} = (\omega/C)\mathbf{e}_r\}$ in spherical polar coordinates, with $\mathbf{e}_r$ the unit radial vector. Because the wave is dispersionless, its phase φ must be conserved along a ray [Eq. (7.28)], so φ must be a function of $Ct - r, \theta$, and ϕ. For the waves to propagate radially, it is essential that $\mathbf{k} = \boldsymbol{\nabla}\varphi$ point very nearly radially, which implies that φ must be a rapidly varying function of $Ct - r$ and a slowly varying one of θ and ϕ. The law of conservation of quanta in this case reduces to the propagation law $d(rA)/dt = 0$ (Ex. 7.6), so rA is also a constant along the ray; we call it $\mathcal{B}$. Putting this all together, we conclude that the sound waves' pressure perturbation $\psi = \delta P$ has the form

$$\psi = \frac{\mathcal{B}(Ct - r, \theta, \phi)}{r} e^{i\varphi(Ct - r, \theta, \phi)}, \quad (7.42)$$

where the phase φ is rapidly varying in $Ct - r$ and slowly varying in the angles, and the amplitude $\mathcal{B}$ is slowly varying in $Ct - r$ and the angles.

FLEXURAL WAVES

As another example of the geometric-optics propagation laws, consider flexural waves on a spacecraft's tapering antenna. The dispersion relation is $\Omega = k^2\sqrt{D/\Lambda}$ [Eq. (7.6)] with $D/\Lambda \propto h^2$, where h is the antenna's thickness in its direction of bend (or the

---

6. This is a general feature of quantum theory; creation and destruction of quanta require imposed oscillations at the high frequency and short wavelength of the waves themselves, or at some submultiple of them (in the case of nonlinear creation and annihilation processes; Chap. 10).

antenna's diameter, if it has a circular cross section); cf. Eq. (12.33). Since Ω is independent of t, as the waves propagate from the spacecraft to the antenna's tip, their frequency ω is conserved [third of Eqs. (7.39)], which implies by the dispersion relation that $k = (D/\Lambda)^{-1/4}\omega^{1/2} \propto h^{-1/2}$; hence the wavelength decreases as $h^{1/2}$. The group velocity is $V_g = 2(D/\Lambda)^{1/4}\omega^{1/2} \propto h^{1/2}$. Since the energy per quantum $\hbar\omega$ is constant, particle conservation implies that the waves' energy must be conserved, which in this 1-dimensional problem means that the energy flowing through a segment of the antenna per unit time must be constant along the antenna. On physical grounds this constant energy flow rate must be proportional to $A^2 V_g h^2$, which means that the amplitude A must increase $\propto h^{-5/4}$ as the flexural waves approach the antenna's end. A qualitatively similar phenomenon is seen in the cracking of a bullwhip (where the speed of the end can become supersonic).

LIGHT THROUGH A LENS AND ALFVÉN WAVES

Figure 7.3 sketches two other examples: light propagating through a lens and Alfvén waves propagating in the magnetosphere of a planet. In Sec. 7.3.6 and the exercises we explore a variety of other applications, but first we describe how the geometric-optics propagation laws can fail (Sec. 7.3.5).

Exercise 7.6 *Derivation and Practice: Quasi-Spherical Solution to Vacuum Scalar Wave Equation*

Derive the quasi-spherical solution (7.42) of the vacuum scalar wave equation $-\partial^2\psi/\partial t^2 + \nabla^2\psi = 0$ from the geometric-optics laws by the procedure sketched in the text.

7.3.5 Relation to Wave Packets; Limitations of the Eikonal Approximation and Geometric Optics

7.3.5

The form $\psi = Ae^{i\varphi}$ of the waves in the eikonal approximation is remarkably general. At some initial moment of time, A and φ can have any form whatsoever, so long as the two-lengthscale constraints are satisfied [A, $\omega \equiv -\partial\varphi/\partial t$, $\mathbf{k} \equiv \boldsymbol{\nabla}\varphi$, and dispersion relation $\Omega(\mathbf{k}; \mathbf{x}, t)$ all vary on lengthscales long compared to $\lambda = 1/k$ and on timescales long compared to $1/\omega$]. For example, ψ could be as nearly planar as is allowed by the inhomogeneities of the dispersion relation. At the other extreme, ψ could be a moderately narrow wave packet, confined initially to a small region of space (though not too small; its size must be large compared to its mean reduced wavelength). In either case, the evolution will be governed by the above propagation laws.

Of course, the eikonal approximation is an approximation. Its propagation laws make errors, though when the two-lengthscale constraints are well satisfied, the errors will be small for sufficiently short propagation times. Wave packets provide an important example. Dispersion (different group velocities for different wave vectors) causes wave packets to spread (disperse) as they propagate; see Ex. 7.2. This spreading

phenomena missed by geometric optics

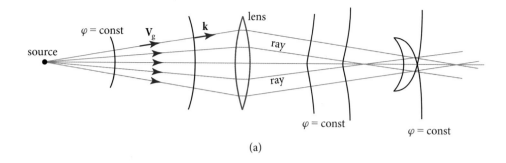

(a)

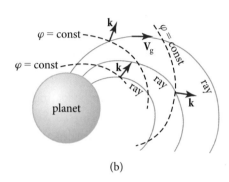

(b)

FIGURE 7.3 (a) The rays and the surfaces of constant phase φ at a fixed time for light passing through a converging lens [dispersion relation $\Omega = ck/\mathrm{n}(\mathbf{x})$, where n is the index of refraction]. In this case the rays (which always point along $\mathbf{V}_\mathrm{g}$) are parallel to the wave vector $\mathbf{k} = \boldsymbol{\nabla}\varphi$ and thus are also parallel to the phase velocity $\mathbf{V}_\mathrm{ph}$, and the waves propagate along the rays with a speed $V_\mathrm{g} = V_\mathrm{ph} = c/\mathrm{n}$ that is independent of wavelength. The strange self-intersecting shape of the last phase front is due to caustics; see Sec. 7.5. (b) The rays and surfaces of constant phase for Alfvén waves in the magnetosphere of a planet [dispersion relation $\Omega = \mathbf{a}(\mathbf{x}) \cdot \mathbf{k}$]. In this case, because $\mathbf{V}_\mathrm{g} = \mathbf{a} \equiv \mathbf{B}/\sqrt{\mu_0\rho}$, the rays are parallel to the magnetic field lines and are not parallel to the wave vector, and the waves propagate along the field lines with speeds V_g that are independent of wavelength; cf. Fig. 7.2c. As a consequence, if some electric discharge excites Alfvén waves on the planetary surface, then they will be observable by a spacecraft when it passes magnetic field lines on which the discharge occurred. As the waves propagate, because $\mathbf{B}$ and ρ are time independent and hence $\partial\Omega/\partial t = 0$, the frequency ω and energy $\hbar\omega$ of each quantum is conserved, and conservation of quanta implies conservation of wave energy. Because the Alfvén speed generally diminishes with increasing distance from the planet, conservation of wave energy typically requires the waves' energy density and amplitude to increase as they climb upward.

is not included in the geometric-optics propagation laws; it is a fundamentally wave-based phenomenon and is lost when one goes to the particle-motion regime. In the limit that the wave packet becomes very large compared to its wavelength or that the packet propagates for only a short time, the spreading is small (Ex. 7.2). This is the geometric-optics regime, and geometric optics ignores the spreading.

Many other wave phenomena are missed by geometric optics. Examples are diffraction (e.g., at a geometric-optics caustic; Secs. 7.5 and 8.6), nonlinear wave-wave coupling (Chaps. 10 and 23, and Sec. 16.3), and parametric amplification of waves by rapid time variations of the medium (Sec. 10.7.3)—which shows up in quantum mechanics as particle production (i.e., a breakdown of the law of conservation of quanta). In Sec. 28.7.1 , we will encounter such particle production in inflationary models of the early universe.

7.3.6 Fermat's Principle

Hamilton's equations of optics allow us to solve for the paths of rays in media that vary both spatially and temporally. When the medium is time independent, the rays $\mathbf{x}(t)$ can be computed from a variational principle due to Fermat. This is the optical analog of the classical dynamics principle of least action,[7] which states that, when a particle moves from one point to another through a time-independent potential (so its energy, the hamiltonian, is conserved), then the path $\mathbf{q}(t)$ that it follows is one that extremizes the action

principle of least action

$$J = \int \mathbf{p} \cdot d\mathbf{q} \tag{7.43}$$

(where $\mathbf{q}$ and $\mathbf{p}$ are the particle's generalized coordinates and momentum), subject to the constraint that the paths have a fixed starting point, a fixed endpoint, and constant energy. The proof (e.g., Goldstein, Poole, and Safko, 2002, Sec. 8.6) carries over directly to optics when we replace the hamiltonian by Ω, $\mathbf{q}$ by $\mathbf{x}$, and $\mathbf{p}$ by $\mathbf{k}$. The resulting Fermat principle, stated with some care, has the following form.

Consider waves whose hamiltonian $\Omega(\mathbf{k}, \mathbf{x})$ is independent of time. Choose an initial location $\mathbf{x}_{\text{initial}}$ and a final location $\mathbf{x}_{\text{final}}$ in space, and consider the rays $\mathbf{x}(t)$ that connect these two points. The rays (usually only one) are those paths that satisfy the variational principle

Fermat's principle

$$\boxed{\delta \int \mathbf{k} \cdot d\mathbf{x} = 0.} \tag{7.44}$$

In this variational principle, $\mathbf{k}$ must be expressed in terms of the trial path $\mathbf{x}(t)$ using Hamilton's equation $dx^j/dt = -\partial\Omega/\partial k_j$; the rate that the trial path is traversed (i.e., the magnitude of the group velocity) must be adjusted to keep Ω constant along the trial path (which means that the total time taken to go from $\mathbf{x}_{\text{initial}}$ to $\mathbf{x}_{\text{final}}$ can differ from one trial path to another). And of course, the trial paths must all begin at $\mathbf{x}_{\text{initial}}$ and end at $\mathbf{x}_{\text{final}}$.

PATH INTEGRALS

Notice that, once a ray has been identified by this action principle, it has $\mathbf{k} = \nabla\varphi$, and therefore the extremal value of the action $\int \mathbf{k} \cdot d\mathbf{x}$ along the ray is equal to the waves'

---

7. This is commonly attributed to Maupertuis, though others, including Leibniz and Euler, understood it earlier or better. This "action" and the rules for its variation are different from those in play in Hamilton's principle.

phase difference $\Delta\varphi$ between $\mathbf{x}_{\mathrm{initial}}$ and $\mathbf{x}_{\mathrm{final}}$. Correspondingly, for any trial path, we can think of the action as a phase difference along that path,

$$\Delta\varphi = \int \mathbf{k} \cdot d\mathbf{x}, \tag{7.45a}$$

and we can think of Fermat's principle as saying that the particle travels along a path of extremal phase difference $\Delta\varphi$. This can be reexpressed in a form closely related to *Feynman's path-integral formulation of quantum mechanics* (Feynman, 1966). We can regard all the trial paths as being followed with equal probability. For each path, we are to construct a probability amplitude $e^{i\Delta\varphi}$, and we must then add together these amplitudes,

$$\sum_{\mathrm{all\ paths}} e^{i\Delta\varphi}, \tag{7.45b}$$

to get the net complex amplitude for quanta associated with the waves to travel from $\mathbf{x}_{\mathrm{initial}}$ to $\mathbf{x}_{\mathrm{final}}$. The contributions from almost all neighboring paths will interfere destructively. The only exceptions are those paths whose neighbors have the same values of $\Delta\varphi$, to first order in the path difference. These are the paths that extremize the action (7.44): they are the wave's rays, the actual paths of the quanta.

SPECIALIZATION TO $\Omega = C(\mathbf{x})k$

Fermat's principle takes on an especially simple form when not only is the hamiltonian $\Omega(\mathbf{k}, \mathbf{x})$ time independent, but it also has the simple dispersion-free form $\Omega = C(\mathbf{x})k$—a form valid for the propagation of light through a time-independent dielectric, and sound waves through a time-independent, inhomogeneous fluid, and electromagnetic or gravitational waves through a time-independent, Newtonian gravitational field (Sec. 7.6). In this $\Omega = C(\mathbf{x})k$ case, the hamiltonian dictates that for each trial path, $\mathbf{k}$ is parallel to $d\mathbf{x}$, and therefore $\mathbf{k} \cdot d\mathbf{x} = k\,ds$, where s is distance along the path. Using the dispersion relation $k = \Omega/C$ and noting that Hamilton's equation $dx^j/dt = \partial\Omega/\partial k_j$ implies $ds/dt = C$ for the rate of traversal of the trial path, we see that $\mathbf{k} \cdot d\mathbf{x} = k\,ds = \Omega\,dt$. Since the trial paths are constrained to have Ω constant, Fermat's principle (7.44) becomes a principle of extremal time: The rays between $\mathbf{x}_{\mathrm{initial}}$ and $\mathbf{x}_{\mathrm{final}}$ are those paths along which

principle of extreme time for dispersionless wave

$$\boxed{\int dt = \int \frac{ds}{C(\mathbf{x})} = \int \frac{\mathrm{n}(\mathbf{x})}{c}ds} \tag{7.46}$$

is extremal. In the last expression we have adopted the convention used for light in a dielectric medium, that $C(\mathbf{x}) = c/\mathrm{n}(\mathbf{x})$, where c is the speed of light in vacuum, and n

index of refraction

is the medium's index of refraction. Since c is constant, the rays are paths of extremal optical path length $\int \mathrm{n}(\mathbf{x})ds$.

We can use Fermat's principle to demonstrate that, if the medium contains no opaque objects, then there will always be at least one ray connecting any two points. This is because there is a lower bound on the optical path between any two points, given by $n_{min}L$, where n_{min} is the lowest value of the refractive index anywhere in the medium, and L is the distance between the two points. This means that for some path the optical path length must be a minimum, and that path is then a ray connecting the two points.

From the principle of extremal time, we can derive the Euler-Lagrange differential equation for the ray. For ease of derivation, we write the action principle in the form

$$\delta \int n(\mathbf{x}) \sqrt{\frac{d\mathbf{x}}{d\mathbf{s}} \cdot \frac{d\mathbf{x}}{d\mathbf{s}}} \, ds, \tag{7.47}$$

where the quantity in the square root is identically one. Performing a variation in the usual manner then gives

$$\boxed{\frac{d}{ds}\left(n\frac{d\mathbf{x}}{ds}\right) = \nabla n, \quad \text{i.e.,} \quad \frac{d}{ds}\left(\frac{1}{C}\frac{d\mathbf{x}}{ds}\right) = \nabla\left(\frac{1}{C}\right).} \tag{7.48}$$

ray equation for
dispersionless wave

This is equivalent to Hamilton's equations for the ray, as one can readily verify using the hamiltonian $\Omega = kc/n$ (Ex. 7.7).

Equation (7.48) is a second-order differential equation requiring two boundary conditions to define a solution. We can either choose these to be the location of the start of the ray and its starting direction, or the start and end of the ray. A simple case arises when the medium is stratified [i.e., when $n = n(z)$, where (x, y, z) are Cartesian coordinates]. Projecting Eq. (7.48) perpendicular to $\mathbf{e}_z$, we discover that $n\,dy/ds$ and $n\,dx/ds$ are constant, which implies

$$\boxed{n \sin \theta = \text{constant},} \tag{7.49}$$

Snell's law

where θ is the angle between the ray and $\mathbf{e}_z$. This is a variant of Snell's law of refraction. Snell's law is just a mathematical statement that the rays are normal to surfaces (wavefronts) on which the eikonal (phase) φ is constant (cf. Fig. 7.4).[8] Snell's law is valid not only when $n(\mathbf{x})$ varies slowly but also when it jumps discontinuously, despite the assumptions underlying geometric optics failing at a discontinuity.

EXERCISES

Exercise 7.7 *Derivation: Hamilton's Equations for Dispersionless Waves;*
Fermat's Principle
Show that Hamilton's equations for the standard dispersionless dispersion relation (7.4) imply the same ray equation (7.48) as we derived using Fermat's principle.

---

8. Another important application of this general principle is to the design of optical instruments, where it is known as the *Abbé condition*. See, e.g., Born and Wolf (1999).

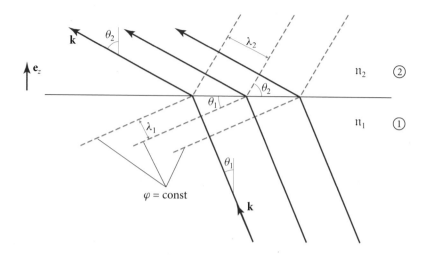

FIGURE 7.4 Illustration of Snell's law of refraction at the interface between two media, for which the refractive indices are n_1 and n_2 (assumed less than n_1). As the wavefronts must be continuous across the interface, simple geometry tells us that $\lambda_1/\sin\theta_1 = \lambda_2/\sin\theta_2$. This and the fact that the wavelengths are inversely proportional to the refractive index, $\lambda_j \propto 1/n_j$, imply that $n_1\sin\theta_1 = n_2\sin\theta_2$, in agreement with Eq. (7.49).

Exercise 7.8 *Example: Self-Focusing Optical Fibers*

Optical fibers in which the refractive index varies with radius are commonly used to transport optical signals. When the diameter of the fiber is many wavelengths, we can use geometric optics. Let the refractive index be

$$n = n_0(1 - \alpha^2 r^2)^{1/2}, \tag{7.50a}$$

where n_0 and α are constants, and r is radial distance from the fiber's axis.

(a) Consider a ray that leaves the axis of the fiber along a direction that makes an angle β to the axis. Solve the ray-transport equation (7.48) to show that the radius of the ray is given by

$$r = \frac{\sin\beta}{\alpha}\left|\sin\left(\frac{\alpha z}{\cos\beta}\right)\right|, \tag{7.50b}$$

where z measures distance along the fiber.

(b) Next consider the propagation time T for a light pulse propagating along the ray with $\beta \ll 1$, down a long length L of fiber. Show that

$$T = \frac{n_0 L}{C}[1 + O(\beta^4)], \tag{7.50c}$$

and comment on the implications of this result for the use of fiber optics for communication.

Exercise 7.9 **Example: Geometric Optics for the Schrödinger Equation*
Consider the nonrelativistic Schrödinger equation for a particle moving in a time-dependent, 3-dimensional potential well:

$$-\frac{\hbar}{i}\frac{\partial \psi}{\partial t} = \left[\frac{1}{2m}\left(\frac{\hbar}{i}\nabla\right)^2 + V(\mathbf{x}, t)\right]\psi. \qquad (7.51)$$

(a) Seek a geometric-optics solution to this equation with the form $\psi = Ae^{iS/\hbar}$, where A and V are assumed to vary on a lengthscale $\mathcal{L}$ and timescale $\mathcal{T}$ long compared to those, $1/k$ and $1/\omega$, on which S varies. Show that the leading-order terms in the two-lengthscale expansion of the Schrödinger equation give the Hamilton-Jacobi equation

$$\frac{\partial S}{\partial t} + \frac{1}{2m}(\nabla S)^2 + V = 0. \qquad (7.52a)$$

Our notation $\varphi \equiv S/\hbar$ for the phase φ of the wave function ψ is motivated by the fact that the geometric-optics limit of quantum mechanics is classical mechanics, and the function $S = \hbar\varphi$ becomes, in that limit, "Hamilton's principal function," which obeys the Hamilton-Jacobi equation (see, e.g., Goldstein, Poole, and Safko, 2002, Chap. 10). [Hint: Use a formal parameter σ to keep track of orders (Box 7.2), and argue that terms proportional to $\hbar^n$ are of order σ^n. This means there must be factors of σ in the Schrödinger equation (7.51) itself.]

(b) From Eq. (7.52a) derive the equation of motion for the rays (which of course is identical to the equation of motion for a wave packet and therefore is also the equation of motion for a classical particle):

$$\frac{d\mathbf{x}}{dt} = \frac{\mathbf{p}}{m}, \qquad \frac{d\mathbf{p}}{dt} = -\nabla V, \qquad (7.52b)$$

where $\mathbf{p} = \nabla S$.

(c) Derive the propagation equation for the wave amplitude A and show that it implies

$$\frac{d|A|^2}{dt} + |A|^2\frac{\nabla \cdot \mathbf{p}}{m} = 0. \qquad (7.52c)$$

Interpret this equation quantum mechanically.

7.4 Paraxial Optics

It is quite common in optics to be concerned with a bundle of rays that are almost parallel (i.e., for which the angle the rays make with some reference ray can be treated as small). This approximation is called *paraxial optics,* and it permits one to linearize the geometric-optics equations and use matrix methods to trace their

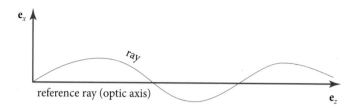

FIGURE 7.5 A reference ray (the z-axis) and an adjacent ray identified by its transverse distances $x(z)$ and $y(z)$, from the reference ray.

rays. The resulting matrix formalism underlies the first-order theory of simple optical instruments (e.g., the telescope and the microscope).

We develop the paraxial optics formalism for waves whose dispersion relation has the simple, time-independent, nondispersive form $\Omega = kc/\mathfrak{n}(\mathbf{x})$. This applies to light in a dielectric medium—the usual application. As we shall see, it also applies to charged particles in a storage ring or electron microscope (Sec. 7.4.2) and to light being lensed by a weak gravitational field (Sec. 7.6).

We restrict ourselves to a situation where there exists a ray that is a straight line, except when it reflects off a mirror or other surface. We choose this as a *reference ray* (also called the *optic axis*) for our formalism, and we orient the z-axis of a Cartesian coordinate system along it (Fig. 7.5). Let the 2-dimensional vector $\mathbf{x}(z)$ be the transverse displacement of some other ray from this reference ray, and denote by $(x, y) = (x_1, x_2)$ the Cartesian components of $\mathbf{x}$.

Under paraxial conditions, $|\mathbf{x}|$ is small compared to the z lengthscales of the propagation, so we can Taylor expand the refractive index $\mathfrak{n}(\mathbf{x}, z)$ in (x_1, x_2):

$$\mathfrak{n}(\mathbf{x}, z) = \mathfrak{n}(0, z) + x_i \mathfrak{n}_{,i}(0, z) + \frac{1}{2} x_i x_j \mathfrak{n}_{,ij}(0, z) + \dots . \tag{7.53a}$$

Here the subscript commas denote partial derivatives with respect to the transverse coordinates, $\mathfrak{n}_{,i} \equiv \partial \mathfrak{n}/\partial x_i$. The linearized form of the ray-propagation equation (7.48) is then given by

$$\frac{d}{dz}\left(\mathfrak{n}(0, z)\frac{dx_i}{dz}\right) = \mathfrak{n}_{,i}(0, z) + x_j \mathfrak{n}_{,ij}(0, z). \tag{7.53b}$$

In order for the reference ray $x_i = 0$ to satisfy this equation, $\mathfrak{n}_{,i}(0, z)$ must vanish, so Eq. (7.53b) becomes a linear, homogeneous, second-order equation for the path of a nearby ray, $\mathbf{x}(z)$:

paraxial ray equation

$$\boxed{\left(\frac{d}{dz}\right)\left(\frac{\mathfrak{n}\,dx_i}{dz}\right) = x_j \mathfrak{n}_{,ij}.} \tag{7.54}$$

Here $\mathfrak{n}$ and $\mathfrak{n}_{,ij}$ are evaluated on the reference ray. It is helpful to regard z as "time" and think of Eq. (7.54) as an equation for the 2-dimensional motion of a particle (the ray) in a quadratic potential well. We can solve Eq. (7.54) given starting values $\mathbf{x}(z')$ and $\dot{\mathbf{x}}(z')$, where the dot denotes differentiation with respect to z, and z' is the starting location. The solution at some later point z is linearly related to the starting values. We can capitalize on this linearity by treating $\{\mathbf{x}(z), \dot{\mathbf{x}}(z)\}$ as a 4-dimensional vector $V_i(z)$, with

$$\boxed{V_1 = x, \quad V_2 = \dot{x}, \quad V_3 = y, \quad \text{and} \quad V_4 = \dot{y},} \tag{7.55a}$$

and embodying the linear transformation [linear solution of Eq. (7.54)] from location z' to location z in a *transfer matrix* $J_{ab}(z, z')$:

$$\boxed{V_a(z) = J_{ab}(z, z') \cdot V_b(z'),} \tag{7.55b}$$

paraxial transfer matrix

where there is an implied sum over the repeated index b. The transfer matrix contains full information about the change of position and direction of all rays that propagate from z' to z. As is always the case for linear systems, the transfer matrix for propagation over a large interval, from z' to z, can be written as the product of the matrices for two subintervals, from z' to z'' and from z'' to z:

$$\boxed{J_{ac}(z, z') = J_{ab}(z, z'') J_{bc}(z'', z').} \tag{7.55c}$$

7.4.1 Axisymmetric, Paraxial Systems: Lenses, Mirrors, Telescopes, Microscopes, and Optical Cavities

7.4.1

If the index of refraction is everywhere axisymmetric, so $\mathfrak{n} = \mathfrak{n}(\sqrt{x^2 + y^2}, z)$, then there is no coupling between the motions of rays along the x and y directions, and the equations of motion along x are identical to those along y. In other words, $J_{11} = J_{33}$, $J_{12} = J_{34}$, $J_{21} = J_{43}$, and $J_{22} = J_{44}$ are the only nonzero components of the transfer matrix. This reduces the dimensionality of the propagation problem from 4 dimensions to 2: V_a can be regarded as either $\{x(z), \dot{x}(z)\}$ or $\{y(z), \dot{y}(z)\}$, and in both cases the 2×2 transfer matrix J_{ab} is the same.

Let us illustrate the paraxial formalism by deriving the transfer matrices of a few simple, axisymmetric optical elements. In our derivations it is helpful conceptually to focus on rays that move in the x-z plane (i.e., that have $y = \dot{y} = 0$). We write the 2-dimensional V_i as a column vector:

axisymmetric transfer matrices

$$\boxed{V_a = \begin{pmatrix} x \\ \dot{x} \end{pmatrix}.} \tag{7.56a}$$

The simplest case is a straight section of length d extending from z' to $z = z' + d$. The components of V will change according to

$$x = x' + \dot{x}'d,$$
$$\dot{x} = \dot{x}',$$

so

$$J_{ab} = \begin{pmatrix} 1 & d \\ 0 & 1 \end{pmatrix} \text{ for a straight section of length } d, \tag{7.56b}$$

where $x' = x(z')$, and so forth. Next, consider a thin lens with focal length f. The usual convention in optics is to give f a positive sign when the lens is converging and a negative sign when diverging. A thin lens gives a deflection to the ray that is linearly proportional to its displacement from the optic axis, but does not change its transverse location. Correspondingly, the transfer matrix in crossing the lens (ignoring its thickness) is

$$J_{ab} = \begin{pmatrix} 1 & 0 \\ -f^{-1} & 1 \end{pmatrix} \text{ for a thin lens with focal length } f. \tag{7.56c}$$

Similarly, a spherical mirror with radius of curvature R (again adopting a positive sign for a converging mirror and a negative sign for a diverging mirror) has a transfer matrix

$$J_{ab} = \begin{pmatrix} 1 & 0 \\ -2R^{-1} & 1 \end{pmatrix} \text{ for a spherical mirror with radius of curvature } R.$$

$$\tag{7.56d}$$

(Recall our convention that z always increases along a ray, even when the ray reflects off a mirror.)

As a simple illustration, we consider rays that leave a point source located a distance u in front of a converging lens of focal length f, and we solve for the ray positions a distance v behind the lens (Fig. 7.6). The total transfer matrix is the transfer matrix (7.56b) for a straight section, multiplied by the product of the lens transfer matrix (7.56c) and a second straight-section transfer matrix:

$$J_{ab} = \begin{pmatrix} 1 & v \\ 0 & 1 \end{pmatrix} \begin{pmatrix} 1 & 0 \\ -f^{-1} & 1 \end{pmatrix} \begin{pmatrix} 1 & u \\ 0 & 1 \end{pmatrix} = \begin{pmatrix} 1 - vf^{-1} & u + v - uvf^{-1} \\ -f^{-1} & 1 - uf^{-1} \end{pmatrix}. \tag{7.57}$$

When the 1-2 element (upper right entry) of this transfer matrix vanishes, the position of the ray after traversing the optical system is independent of the starting direction. In other words, rays from the point source form a point image. When this happens, the planes containing the source and the image are said to be conjugate. The condition for this to occur is

$$\frac{1}{u} + \frac{1}{v} = \frac{1}{f}. \tag{7.58}$$

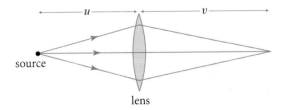

FIGURE 7.6 Simple converging lens used to illustrate the use of transfer matrices. The total transfer matrix is formed by taking the product of the straight-section transfer matrix with the lens matrix and another straight-section matrix.

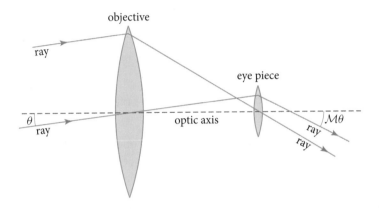

FIGURE 7.7 Simple refracting telescope. By convention $\theta > 0$ and $\mathcal{M}\theta < 0$, so the image is inverted.

This is the standard thin-lens equation. The linear magnification of the image is given by $\mathcal{M} = J_{11} = 1 - v/f$, that is,

$$\mathcal{M} = -\frac{v}{u}, \tag{7.59}$$

where the negative sign means that the image is inverted. Note that, if a ray is reversed in direction, it remains a ray, but with the source and image planes interchanged; u and v are exchanged, Eq. (7.58) is unaffected, and the magnification (7.59) is inverted: $\mathcal{M} \to 1/\mathcal{M}$.

Exercise 7.10 *Problem: Matrix Optics for a Simple Refracting Telescope*
Consider a simple refracting telescope (Fig. 7.7) that comprises two converging lenses, the *objective* and the *eyepiece*. This telescope takes parallel rays of light from distant stars, which make an angle $\theta \ll 1$ with the optic axis, and converts them into parallel rays making a much larger angle $\mathcal{M}\theta$. Here $\mathcal{M}$ is the magnification with $\mathcal{M}$ negative,

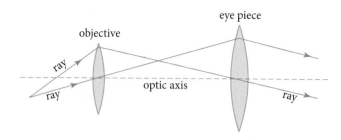

FIGURE 7.8 Simple microscope.

$|\mathcal{M}| \gg 1$, and $|\mathcal{M}\theta| \ll 1$. (The parallel output rays are then focused by the lens of a human's eye, to a point on the eye's retina.)

(a) Use matrix methods to investigate how the output rays depend on the separation of the two lenses, and hence find the condition that the output rays are parallel when the input rays are parallel.

(b) How does the magnification $\mathcal{M}$ depend on the ratio of the focal lengths of the two lenses?

(c) If, instead of looking through the telescope with one's eye, one wants to record the stars' image on a photographic plate or CCD, how should the optics be changed?

Exercise 7.11 *Problem: Matrix Optics for a Simple Microscope*
A microscope takes light rays from a point on a microscopic object, very near the optic axis, and transforms them into parallel light rays that will be focused by a human eye's lens onto the eye's retina (Fig. 7.8). Use matrix methods to explore the operation of such a microscope. A single lens (magnifying glass) could do the same job (rays from a point converted to parallel rays). Why does a microscope need two lenses? What focal lengths and lens separations are appropriate for the eye to resolve a bacterium 100 μm in size?

Exercise 7.12 *Example: Optical Cavity—Rays Bouncing between Two Mirrors*
Consider two spherical mirrors, each with radius of curvature R, separated by distance d so as to form an optical cavity (Fig. 7.9). A laser beam bounces back and forth between the two mirrors. The center of the beam travels along a geometric-optics ray. (We study such beams, including their diffractive behavior, in Sec. 8.5.5.)

(a) Show, using matrix methods, that the central ray hits one of the mirrors (either one) at successive locations $\mathbf{x}_1, \mathbf{x}_2, \mathbf{x}_3, \ldots$ (where $\mathbf{x} \equiv (x, y)$ is a 2-dimensional vector in the plane perpendicular to the optic axis), which satisfy the difference equation

$$\mathbf{x}_{k+2} - 2b\mathbf{x}_{k+1} + \mathbf{x}_k = 0, \tag{7.60a}$$

FIGURE 7.9 An optical cavity formed by two mirrors, and a light beam bouncing back and forth inside it.

where

$$b = 1 - \frac{4d}{R} + \frac{2d^2}{R^2}. \tag{7.60b}$$

Explain why this is a difference-equation analog of the simple-harmonic-oscillator equation.

(b) Show that this difference equation has the general solution

$$\mathbf{x}_k = \mathbf{A} \cos(k \cos^{-1} b) + \mathbf{B} \sin(k \cos^{-1} b). \tag{7.60c}$$

Obviously, $\mathbf{A}$ is the transverse position $\mathbf{x}_0$ of the ray at its 0th bounce. The ray's 0th position $\mathbf{x}_0$ and its 0th direction of motion $\dot{\mathbf{x}}_0$ together determine $\mathbf{B}$.

(c) Show that if $0 \leq d \leq 2R$, the mirror system is stable. In other words, all rays oscillate about the optic axis. Similarly, show that if $d > 2R$, the mirror system is unstable and the rays diverge from the optic axis.

(d) For an appropriate choice of initial conditions $\mathbf{x}_0$ and $\dot{\mathbf{x}}_0$, the laser beam's successive spots on the mirror lie on a circle centered on the optic axis. When operated in this manner, the cavity is called a *Harriet delay line*. How must d/R be chosen so that the spots have an angular step size θ? (There are two possible choices.)

7.4.2 Converging Magnetic Lens for Charged Particle Beam

Since geometric optics is the same as particle dynamics, matrix equations can be used to describe paraxial motions of electrons or ions in a storage ring. (Note, however, that the hamiltonian for such particles is dispersive, since it does not depend linearly on the particle momentum, and so for our simple matrix formalism to be valid, we must confine attention to a monoenergetic beam of particles.)

The simplest practical lens for charged particles is a quadrupolar magnet. Quadrupolar magnetic fields are used to guide particles around storage rings. If we orient our axes appropriately, the magnet's magnetic field can be expressed in the form

$$\mathbf{B} = \frac{B_0}{r_0}(y\mathbf{e}_x + x\mathbf{e}_y) \quad \text{independent of } z \text{ within the lens} \tag{7.61}$$

quadrupolar magnetic field

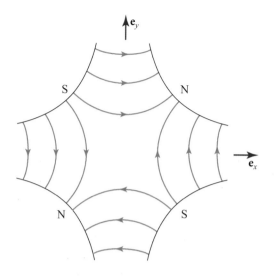

FIGURE 7.10 Quadrupolar magnetic lens. The magnetic field lines lie in a plane perpendicular to the optic axis. Positively charged particles moving along $\mathbf{e}_z$ converge when $y = 0$ and diverge when $x = 0$.

(Fig. 7.10). Particles traversing this magnetic field will be subjected to a Lorentz force that curves their trajectories. In the paraxial approximation, a particle's coordinates satisfy the two differential equations

$$\ddot{x} = -\frac{x}{\lambda^2}, \qquad \ddot{y} = \frac{y}{\lambda^2},$$

(7.62a)

where the dots (as above) mean $d/dz = v^{-1}d/dt$, and

$$\lambda = \left(\frac{p r_0}{q B_0}\right)^{1/2}$$

(7.62b)

[cf. Eq. (7.61)], with q the particle's charge (assumed positive) and p its momentum. The motions in the x and y directions are decoupled. It is convenient in this case to work with two 2-dimensional vectors, $\{V_{x1}, V_{x2}\} \equiv \{x, \dot{x}\}$ and $\{V_{y1}, V_{y2}\} = \{y, \dot{y}\}$. From the elementary solutions to the equations of motion (7.62a), we infer that the transfer matrices from the magnet's entrance to its exit are $J_{x\,ab}$, $J_{y\,ab}$, where

transfer matrices for quadrupolar magnetic lens

$$J_{x\,ab} = \begin{pmatrix} \cos\phi & \lambda\,\sin\phi \\ -\lambda^{-1}\,\sin\phi & \cos\phi \end{pmatrix},$$

(7.63a)

$$J_{y\,ab} = \begin{pmatrix} \cosh\phi & \lambda\,\sinh\phi \\ \lambda^{-1}\,\sinh\phi & \cosh\phi \end{pmatrix},$$

(7.63b)

and

$$\phi = L/\lambda,$$

(7.63c)

with L the distance from entrance to exit (i.e., the lens thickness).

The matrices $J_{x\,ab}$ and $J_{y\,ab}$ can be decomposed as follows:

$$J_{x\,ab} = \begin{pmatrix} 1 & \lambda \tan \phi/2 \\ 0 & 1 \end{pmatrix} \begin{pmatrix} 1 & 0 \\ -\lambda^{-1} \sin \phi & 1 \end{pmatrix} \begin{pmatrix} 1 & \lambda \tan \phi/2 \\ 0 & 1 \end{pmatrix} \quad (7.63\text{d})$$

$$J_{y\,ab} = \begin{pmatrix} 1 & \lambda \tanh \phi/2 \\ 0 & 1 \end{pmatrix} \begin{pmatrix} 1 & 0 \\ \lambda^{-1} \sinh \phi & 1 \end{pmatrix} \begin{pmatrix} 1 & \lambda \tanh \phi/2 \\ 0 & 1 \end{pmatrix} \quad (7.63\text{e})$$

Comparing with Eqs. (7.56b) and (7.56c), we see that the action of a single magnet is equivalent to the action of a straight section, followed by a thin lens, followed by another straight section. Unfortunately, if the lens is focusing in the x direction, it must be defocusing in the y direction and vice versa. However, we can construct a lens that is focusing along both directions by combining two magnets that have opposite polarity but the same focusing strength $\phi = L/\lambda$.

combining two magnets to make a converging lens

Consider first the particles' motion in the x direction. Let

$$f_+ = \lambda/\sin \phi \quad \text{and} \quad f_- = -\lambda/\sinh \phi \quad (7.64)$$

be the equivalent focal lengths of the first converging lens and the second diverging lens. If we separate the magnets by a distance s, this must be added to the two effective lengths of the two magnets to give an equivalent separation of $d = \lambda \tan(\phi/2) + s + \lambda \tanh(\phi/2)$ for the two equivalent thin lenses. The combined transfer matrix for the two thin lenses separated by this distance d is then

$$\begin{pmatrix} 1 & 0 \\ -f_-^{-1} & 1 \end{pmatrix} \begin{pmatrix} 1 & d \\ 0 & 1 \end{pmatrix} \begin{pmatrix} 1 & 0 \\ -f_+^{-1} & 1 \end{pmatrix} = \begin{pmatrix} 1 - df_+^{-1} & d \\ -f_*^{-1} & 1 - df_-^{-1} \end{pmatrix}, \quad (7.65\text{a})$$

transfer matrix for converging magnetic lens

where

$$\frac{1}{f_*} = \frac{1}{f_-} + \frac{1}{f_+} - \frac{d}{f_- f_+} = \frac{\sin \phi}{\lambda} - \frac{\sinh \phi}{\lambda} + \frac{d \sin \phi \sinh \phi}{\lambda^2}. \quad (7.65\text{b})$$

If we assume that $\phi \ll 1$ and $s \ll L$, then we can expand as a Taylor series in ϕ to obtain

$$f_* \simeq \frac{3\lambda}{2\phi^3} = \frac{3\lambda^4}{2L^3}. \quad (7.66)$$

The effective focal length f_* of the combined magnets is positive, and so the lens has a net focusing effect. From the symmetry of Eq. (7.65b) under interchange of f_+ and f_-, it should be clear that f_* is independent of the order in which the magnets are encountered. Therefore, if we were to repeat the calculation for the motion in the y direction, we would get the same focusing effect. (The diagonal elements of the transfer matrix are interchanged, but as they are both close to unity, this difference is rather small.)

The combination of two quadrupole lenses of opposite polarity can therefore imitate the action of a converging lens. Combinations of magnets like this are used to collimate particle beams in storage rings, particle accelerators, and electron microscopes.

7.5 Catastrophe Optics T2

7.5.1 Image Formation T2

CAUSTICS

Many simple optical instruments are carefully made to form point images from point sources. However, naturally occurring optical systems, and indeed precision optical instruments when examined in fine detail, bring light to a focus not at a point, but instead on a 2-dimensional surface—an envelope formed by the rays—called a *caustic*. Caustics are often seen in everyday life. For example, when bright sunlight is reflected by the inside of an empty coffee mug some of the rays are reflected *specularly* (angle of incidence equals angle of reflection) and some of the rays are reflected *diffusely* (in all directions due to surface irregularity and multiple reflections and refractions beneath the surface). The specular reflection by the walls—a cylindrical mirror— forms a caustic surface. The intersection of this surface with the bottom forms caustic lines that can be seen in diffuse reflection.[9] These caustic lines are observed to meet in a point. When the optical surfaces are quite irregular (e.g., the water surface in a swimming pool[10] or the type of glass used in bathrooms), then a *caustic network* forms. Caustic lines and points are seen, just the same as with the mug (Fig. 7.11).

What may be surprising is that caustics like these, formed under quite general conditions, can be classified into a rather small number of types, called *catastrophes*, possessing generic properties and scaling laws (Thom, 1994). The scaling laws are reminiscent of the renormalization group discussed in Sec. 5.8.3. Although we focus on catastrophes in the context of optics (e.g., Berry and Upstill, 1980), where they are caustics, the phenomenon is quite general and crops up in other subfields of physics, especially dynamics (e.g., Arnol'd, 1992) and thermodynamics (e.g., Ex. 7.16). It has also been invoked, often quite controversially, in fields outside physics (e.g., Poston and Stewart, 2012). Catastrophes can be found whenever we have a physical system whose states are determined by extremizing a function, such as energy. Our treatment will be quite heuristic, but the subject does have a formal mathematical foundation that connects it to bifurcations and *Morse theory* (see Sec. 11.6; see also, e.g., Petters et al., 2001).

STATE VARIABLES AND CONTROL PARAMETERS

Let us start with a specific, simple example. Suppose that there is a distant *source* S and a *detector* D separated by free space. If we consider all the paths from S to

---

9. The curve that is formed is called a "nephroid."
10. The optics is quite complicated. Some rays from the Sun are reflected specularly by the surface of the water, creating multiple images of the Sun. As the Sun is half a degree in diameter, these produce thickened caustic lines. Some rays are refracted by the water, forming caustic surfaces that are intersected by the bottom of the pool to form a caustic pattern. Some light from this pattern is reflected diffusely before being refracted a second time on the surface of the water and ultimately detected by a retina or a CCD. Other rays are reflected multiple times.

caustic

catastrophes

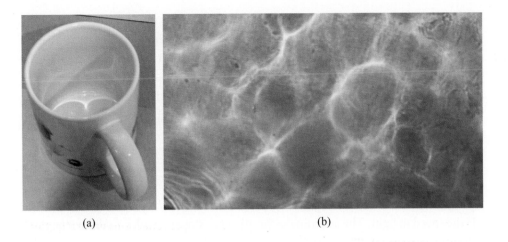

(a)   (b)

FIGURE 7.11 Photographs of caustics. (a) Simple caustic pattern formed by a coffee mug. (b) Caustic network formed in a swimming pool. The generic structure of these patterns comprises *fold* lines meeting at *cusp* points.

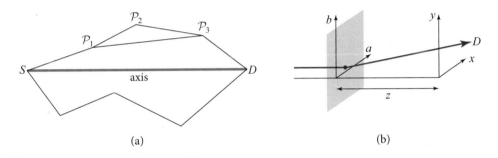

(a)   (b)

FIGURE 7.12 (a) Alternative paths make up a sequence of straight segments from a source S to a detector D. The path S-P_1-P_2-P_3-D can be simplified to a shorter path S-P_1-P_3-D, and this process can be continued until we have the minimum number of segments needed to exhibit the catastrophe. The (true) ray, with the smallest phase difference, is the axis S-D. (b) A single path from a distant source intersecting a screen at $\{a, b\}$ and ending at detector D with coordinates $\{x, y, z\}$.

D there is a single extremum—a minimum—in the phase difference, $\Delta\varphi = \omega t = \int \mathbf{k} \cdot d\mathbf{x}$. By Fermat's principle, this is the (true) ray—the *axis*—connecting the two points. There are an infinite number of alternative paths that could be defined by an infinite set of parameters, but wherever else the rays go, the phase difference is larger. Take one of these alternative paths connecting S and D and break it down into a sequence of connected segments (Fig. 7.12a). We can imagine replacing two successive segments with a single segment connecting their endpoints. This will reduce the phase difference. The operation can be repeated until we are left with the minimum number of segments, specified by the minimum number of necessary variables that we need to exhibit the catastrophe. The order in which we do this does not matter, and the final variables characterizing the path can be chosen for convenience. These variables are known as *state variables*.

state variables

Next, introduce a screen perpendicular to the S–D axis and close to D (Fig. 7.12b). Consider a path from S, nearly parallel to the axis and intersecting the screen at a point with Cartesian coordinates $\{a, b\}$ measured from the axis. There let it be deflected toward D. In this section and the next, introduce the *delay* $t \equiv \Delta\varphi/\omega$, subtracting off the constant travel time along the axis in the absence of the screen, to measure the phase. The additional geometric delay associated with this ray is given approximately by

$$t_{\text{geo}} = \frac{a^2 + b^2}{2zc}, \tag{7.67}$$

control parameters

where $z \gg \{a, b\}$ measures the distance from the screen to D, parallel to the axis, and c is the speed of light. The coordinates $\{a, b\}$ act as state variables, and the true ray is determined by differentiating with respect to them. Next, move D off the axis and give it Cartesian coordinates $\{x, y, z\}$ with the x-y plane parallel to the a-b plane, and the transverse coordinates measured from the original axis. As these coordinates specify one of the endpoints, they do not enter into the variation that determines the true ray, but they do change t_{geo} to $[(a - x)^2 + (b - y)^2]/(2zc)$. These $\{x, y, z\}$ parameters are examples of *control parameters*. In general, the number of control parameters that we use is also the minimum needed to exhibit the catastrophe, and the choice is usually determined by algebraic convenience.

FOLD CATASTROPHE

Now replace the screen with a thin lens of refractive index $\mathfrak{n}$ and thickness $w(a, b)$. This introduces an additional contribution to the delay, $t_{\text{lens}} = (\mathfrak{n} - 1)w/c$. The true ray will be fixed by the variation of the sum of the geometric and lens delays with respect to a and b plus any additional state variables that are needed. Suppose that the lens is cylindrical so rays are bent only in the x direction and one state variable, a, suffices. Let us use an analytically tractable example, $t_{\text{lens}} = s^2(1 - 2a^2/s^2 + a^4/s^4)/(4fc)$, for $|a| < s$, where $f \gg s$ is the focal length (Fig. 7.13). Place the detector D on the axis with $z < f$. The delay along a path is:

$$t \equiv t_{\text{geo}} + t_{\text{lens}} = \frac{a^2}{2fzc}\left(f - z + \frac{a^2z}{2s^2}\right), \tag{7.68}$$

dropping a constant. This leaves the single minimum and the true ray at $a = 0$.

Now displace D perpendicular to the axis a distance x with z fixed. t_{geo} becomes $(a - x)^2/(2zc)$, and the true ray will be parameterized by the single real value of a that minimizes t and therefore solves

$$x = \frac{a}{f}\left(f - z + z\frac{a^2}{s^2}\right). \tag{7.69}$$

The first two terms, $f - z$, represent a perfect thin lens [cf. J_{11} in Eq. (7.57)]; the third represents an imperfection.

A human eye at D [coordinates (x, y, z)] focuses the ray through D and adjacent rays onto its retina, producing there a point image of the point source at S. The

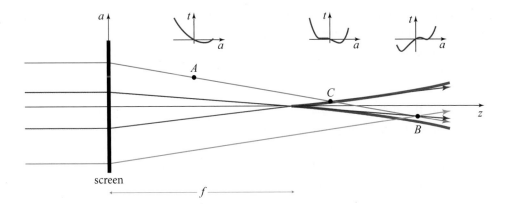

screen

f

FIGURE 7.13 Light from a distant source is normally incident on a thin, phase-changing lens. The phase change depends solely on the distance a from the axis. The delay t along paths encountering a detector D located at (x, z) can be calculated using an equation such as Eq. (7.68). The true rays are located where t is extremized. The envelope created by these rays comprises two fold caustic curves (red) that meet at a cusp point. When D lies outside the caustic, for example at A, there is only one true ray (cyan) where t has its single minimum. When D lies inside the caustic, for example at point B, there are three true rays associated with two minima (orange, cyan) and one maximum (purple) of t. When D lies on the caustic, for example at C, there is one minimum (cyan) and a point of inflection (green). The magnification is formally infinite on the caustic. The cusp at the end of the caustic is the focus, a distance f from the screen.

power $P = d\mathcal{E}/dt$ in that image is the energy flux at D times the area inside the eye's iris, so the image power is proportional to the energy flux. Since the energy flux is proportional to the square of the field amplitude A^2 and $A \propto 1/\sqrt{\mathcal{A}}$, where $\mathcal{A}$ is the area of a bundle of rays [Eq. (7.35d)], the image power is $P \propto 1/\mathcal{A}$.

Consider a rectangular ray bundle that passes through $\{a, b\}$ on the screen with edges da and db, and area $\mathcal{A}_{\text{screen}} = dadb$. The bundle's area when it arrives at D is $\mathcal{A}_D = dxdy$. Because the lens is cylindrical, $dy = db$, so the ratio of power seen by an eye at D to that seen by an eye at the screen (i.e., the lens's *magnification*) is $\mathcal{M} = dP_D/dP_{\text{screen}} = \mathcal{A}_{\text{screen}}/\mathcal{A}_D = da/dx$. Using Eq. (7.69), we find

$$\mathcal{H} \equiv \mathcal{M}^{-1} = \frac{dx}{da} = zc\left(\frac{d^2t}{da^2}\right) = \left(\frac{f-z}{f} + 3\frac{a^2z}{s^2f}\right). \tag{7.70}$$

If the curvature $\mathcal{H}$ of the delay in the vicinity of a true ray is decreased, the magnification is increased. When $z < f$, the curvature is positive. When $z > f$ the curvature for $a = x = 0$ is negative, and the magnification is $\mathcal{M} = -f/(z - f)$, corresponding to an inverted image. However, there are now two additional images with $a = \pm z^{-1/2}(z - f)^{1/2}s$ at minima with associated magnifications $\mathcal{M} = \frac{1}{2}f(z-f)^{-1}$.

Next move the detector D farther away from the axis. A maximum and a minimum in t will become a point of inflection and, equivalently, two of the images will merge when $a = a_f = \pm 3^{-1/2}z^{-1/2}(z - f)^{1/2}s$ or

$$x = x_f \equiv \mp 2f^{-1}z^{-1/2}(z - f)^{3/2}s. \tag{7.71}$$

This is the location of the caustic; in 3-dimensional space it is the surface shown in the first panel of Fig. 7.15 below.

The magnification of the two images will diverge at the caustic and, expanding $\mathcal{H}$ to linear order about zero, we find that $\mathcal{M} = \pm 2^{-1}3^{-1/4}f^{1/2}s^{1/2}z^{-1/4}(z-f)^{-1/4}|x - x_f|^{-1/2}$ for each image. However, when $|x| > x_f$, the two images vanish. This abrupt change in the optics—two point images becoming infinitely magnified and then vanishing as the detector is moved—is an example of a *fold catastrophe* occurring at a caustic. (Note that the algebraic sum of the two magnifications is zero in the limit.)

It should be pointed out that the divergence of the magnification does not happen in practice for two reasons. The first is that a point source is only an idealization, and if we allow the source to have finite size, different parts will produce caustics at slightly different locations. The second is that geometric optics, on which our analysis is based, pretends that the wavelength of light is vanishingly small. In actuality, the wavelength is always nonzero, and near a caustic its finiteness leads to diffraction effects, which also limit the magnification to a finite value (Sec. 8.6).

Although we have examined one specific and stylized example, the algebraic details can be worked out for any configuration governed by geometric optics. However, they are less important than the scaling laws—for example, $\mathcal{M} \propto |x - x_f|^{-1/2}$—that become increasingly accurate as the catastrophe (caustic) is approached. For this reason, catastrophes are commonly given a *standard form* chosen to exhibit these features and only valid very close to the catastrophe. We discuss this here just in the context of geometrical optics, but the basic scalings are useful in other physics applications (see, e.g., Ex. 7.16).

First we measure the state variables $a, b, \ldots$ in units of some appropriate scale. Next we do likewise for the control parameters, $x, y, \ldots$. We call the new state variables and control parameters $\tilde{a}, \tilde{b}, \ldots$ and $\tilde{x}, \tilde{y} \ldots$, respectively. We then Taylor expand the delay about the catastrophe. In the case of the fold, we want to be able to find up to two extrema. This requires a cubic equation in $\tilde{a}$. The constant is clearly irrelevant, and an overall multiplying factor will not change the scalings. We are also free to change the origin of $\tilde{a}$, allowing us to drop either the linear or the quadratic term (we choose the latter, so that the coefficient is linearly related to x). If we adjust the scaled delay in the vicinity of a fold catastrophe, Eq. (7.68) can be written in the standard form:

$$\tilde{t}_{\text{fold}} = \frac{1}{3}\tilde{a}^3 - \tilde{x}\tilde{a}, \tag{7.72}$$

where the coefficients are chosen for algebraic convenience. The maximum number of rays involved in the catastrophe is two, and the number of control parameters required is one, which we can think of as being used to adjust the difference in $\tilde{t}$ between two stationary points. The scaled magnifications are now given by $\widetilde{\mathcal{M}} \equiv (d\tilde{x}/d\tilde{a})^{-1} = \pm\frac{1}{2}\tilde{x}^{-1/2}$, and the combined, scaled magnification is $\widetilde{\mathcal{M}} = \tilde{x}^{-1/2}$ for $\tilde{x} > 0$.

CUSP CATASTROPHE

So far, we have only allowed D to move perpendicular to the axis along x. Now move it along the axis toward the screen. We find that x_f decreases with decreasing z until it vanishes at $z = f$. At this point, the central maximum in t merges simultaneously with both minima, leaving a single image. This is an example of a *cusp catastrophe*. Working in 1-dimensional state-variable space with two control parameters and applying the same arguments as we just used with the fold, the standard form for the cusp can be written as

$$\tilde{t}_{\text{cusp}} = \frac{1}{4}\tilde{a}^4 - \frac{1}{2}\tilde{z}\tilde{a}^2 - \tilde{x}\tilde{a}. \tag{7.73}$$

for cusp catastrophe

The parameter $\tilde{x}$ is still associated with a transverse displacement of D, and we can quickly persuade ourselves that $\tilde{z} \propto z - f$ by inspecting the quadratic term in Eq. (7.68).

The cusp then describes a transition between one and three images, one of which must be inverted with respect to the other two. The location of the image for a given $\tilde{a}$ and $\tilde{z}$ is

$$\tilde{x} = \tilde{a}^3 - \tilde{z}\tilde{a}. \tag{7.74}$$

Conversely, for a given $\tilde{x}$ and $\tilde{z}$, there are one or three real solutions for $\tilde{a}(\tilde{x}, \tilde{z})$ and one or three images. The equation satisfied by the fold lines where the transition occurs is

$$\tilde{x} = \pm\frac{2}{3^{3/2}}\tilde{z}^{3/2}. \tag{7.75}$$

These are the two branches of a semi-cubical parabola (the caustic surface in 3 dimensions depicted in the second panel of Fig. 7.15 below), and they meet at the cusp catastrophe where $\tilde{x} = \tilde{z} = 0$.

The scaled magnification at the cusp is

$$\widetilde{\mathcal{M}}(\tilde{x}, \tilde{z}) = \left(\frac{\partial \tilde{x}}{\partial \tilde{a}}\right)_{\tilde{z}}^{-1} = [3\tilde{a}(\tilde{x}, \tilde{z})^2 - z]^{-1}. \tag{7.76}$$

SWALLOWTAIL CATASTROPHE

Now let the rays propagate in 3 dimensions, so that there are three control variables, x, y, and z, where the y-axis is perpendicular to the x- and z-axes. The fold, which was a point in 1 dimension and a line in 2, becomes a surface in 3 dimensions, and the point cusp in 2 dimensions becomes a line in 3 (see Fig. 7.15 below). If there is still only one state variable, then $\tilde{t}$ should be a quintic with up to four extrema. (In general, a catastrophe involving as many as N images requires $N - 1$ control parameters for its full description. These parameters can be thought of as independently changing the relative values of $\tilde{t}$ at the extrema.) The resulting, four-image catastrophe is called a *swallowtail*. (In practice, this catastrophe only arises when there are two state variables, and additional images are always present. However, these are not involved in the catastrophe.) Again following our procedure, we can write the standard form of the swallowtail catastrophe as

for swallowtail catastrophe

$$\tilde{t}_{\text{swallowtail}} = \frac{1}{5}\tilde{a}^5 - \frac{1}{3}\tilde{z}\tilde{a}^3 - \frac{1}{2}\tilde{y}\tilde{a}^2 - \tilde{x}\tilde{a}. \tag{7.77}$$

There are two cusp lines in the half-space $\tilde{z} > 0$, and these meet at the catastrophe where $\tilde{x} = \tilde{y} = \tilde{z} = 0$ (see Fig. 7.15 below). The relationship between $\tilde{x}$, $\tilde{y}$, and $\tilde{z}$ and x, y, and z in this or any other example is not simple, and so the variation of the magnification in the vicinity of the swallowtail catastrophe depends on the details.

HYPERBOLIC UMBILIC CATASTROPHE

Next increase the number of essential state variables to two. We can choose these to be $\tilde{a}$ and $\tilde{b}$. To see what is possible, sketch contours of Δt in the $\tilde{a}$-$\tilde{b}$ plane for fixed values of the control variables. The true rays will be associated with maxima, minima, or saddle points, and each distinct catastrophe corresponds to a different way to nest the contours (Fig. 7.14). The properties of the fold, cusp, and swallowtail are essentially unchanged by the extra dimension. We say that they are *structurally stable*. However, a little geometric experimentation uncovers a genuinely 2-dimensional nesting. The *hyperbolic umbilic* catastrophe has two saddles, one maximum and one minimum. Further algebraic experiment produces a standard form:

for hyperbolic umbilic catastrophe

$$\tilde{t} = \frac{1}{3}(\tilde{a}^3 + \tilde{b}^3) - \tilde{z}\tilde{a}\tilde{b} - \tilde{x}\tilde{a} - \tilde{y}\tilde{b}. \tag{7.78}$$

This catastrophe can be exhibited by a simple generalization of our example. We replace the cylindrical lens described by Eq. (7.68) with a nearly circular lens where the focal length f_a for rays in the a-z plane differs from the focal length f_b for rays in the b-z plane.

$$t = \frac{a^2}{2f_a zc}\left(f_a - z + \frac{a^2 z}{2s_a^2}\right) + \frac{b^2}{2f_b zc}\left(f_b - z + \frac{b^2 z}{2s_b^2}\right). \tag{7.79}$$

astigmatism

This is an example of *astigmatism*. A pair of fold surfaces is associated with each of these foci. These surfaces can cross, and when this happens a cusp line associated with one fold surface can transfer onto the other fold surface. The point where this happens is the hyperbolic umbilic catastrophe.

This example also allows us to illustrate a simple feature of magnification. When the source and the detector are both treated as 2-dimensional, then we generalize the curvature to the *Hessian* matrix

magnification matrix

$$\widetilde{\mathcal{H}} = \widetilde{\mathcal{M}}^{-1} = \begin{pmatrix} \frac{\partial \tilde{x}}{\partial \tilde{a}} & \frac{\partial \tilde{y}}{\partial \tilde{a}} \\ \frac{\partial \tilde{x}}{\partial \tilde{b}} & \frac{\partial \tilde{y}}{\partial \tilde{b}} \end{pmatrix}. \tag{7.80}$$

The magnification matrix $\widetilde{\mathcal{M}}$, which describes the mapping from the source plane to the image plane, is simply the inverse of $\widetilde{\mathcal{H}}$. The four matrix elements also describe the deformation of the image. As we describe in more detail when discussing elastostatics (Sec. 11.2.2), the antisymmetric part of $\widetilde{\mathcal{M}}$ describes the rotation of the image, and

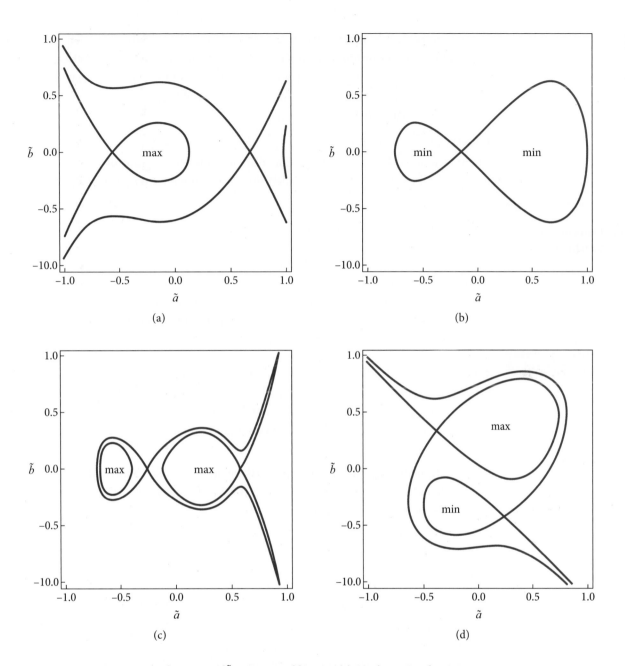

(a)

(b)

(c)

(d)

FIGURE 7.14 Distinct nestings of contours of $\tilde{t}$ in state-variable space. (a) A 1-dimensional arrangement of two saddle points and a maximum in the vicinity of a cusp. The locations of the extrema and their curvatures change as the control parameters change. (b) A cusp formed by two minima and a saddle. Although the nestings look different in 2 dimensions, this is essentially the same catastrophe when considered in 1 dimension, which is all that is necessary to determine its salient properties. These are the only contour nestings possible with two state variables and three extrema (or rays). (c) When we increase the number of extrema to four, two more nestings are possible. The swallowtail catastrophe is essentially a cusp with an additional extremum added to the end, requiring three control parameters to express. It, too, is essentially 1-dimensional. (d) The hyperbolic umbilic catastrophe is essentially 2-dimensional and is associated with a maximum, a minimum, and two saddles. A distinct nesting of contours with three saddle points and one extremum occurs in the elliptic umbilic catastrophe (Ex. 7.13).

the symmetric part its magnification and stretching or *shear*. Both eigenvalues of $\widetilde{\mathcal{M}}$ are positive at a minimum, and the image is a distorted version of the source. At a saddle, one eigenvalue is positive, the other negative, and the image is inverted; at a maximum, they are both negative, and the image is doubly inverted so that it appears to have been rotated through a half-turn.

ELLIPTIC UMBILIC CATASTROPHE

There is a second standard form that can describe the nesting of contours just discussed—a distinct catastrophe called the *elliptic umbilic catastrophe* (Ex. 7.13b):

standard form for elliptic umbilic catastrophe

$$\tilde{t} = \frac{1}{3}\tilde{a}^3 - \tilde{a}\tilde{b}^2 - \tilde{z}(\tilde{a}^2 + \tilde{b}^2) - \tilde{x}\tilde{a} - \tilde{y}\tilde{b}. \tag{7.81}$$

The caustic surfaces in three dimensions $(\tilde{x}, \tilde{y}, \tilde{z})$ for the five elementary catastrophes discussed here are shown in Fig. 7.15. Additional types of catastrophe are found with more control parameters, for example, time (e.g., Poston and Stewart, 2012). This is relevant, for example, to the twinkling of starlight in the geometric-optics limit.

EXERCISES

Exercise 7.13 *Derivation and Problem: Cusps and Elliptic Umbilics* T2

(a) Work through the derivation of Eq. (7.73) for the scaled time delay in the vicinity of the cusp caustic for our simple example [Eq. (7.68)], with the aid of a suitable change of variables (Goodman, Romani, Blandford, and Narayan, 1987, Appendix B).

(b) Sketch the nesting of the contours for the elliptic umbilic catastrophe as shown for the other four catastrophes in Fig. 7.14. Verify that Eq. (7.81) describes this catastrophe.

Exercise 7.14 *Problem: Cusp Scaling Relations* T2

Consider a cusp catastrophe created by a screen as in the example and described by a standard cusp potential, Eq. (7.73). Suppose that a detector lies between the folds, so that there are three images of a single point source with state variables $\tilde{a}_i$.

(a) Explain how, in principle, it is possible to determine $\tilde{a}$ for a single image by measurements at D.

(b) Make a 3-dimensional plot of the location of the image(s) in $\tilde{a}$-$\tilde{x}$-$\tilde{y}$ space and explain why the names "fold" and "cusp" were chosen.

(c) Prove as many as you can of the following scaling relations, valid in the limit as the cusp catastrophe is approached:

$$\sum_{i=1}^{3} \tilde{a}_i = 0, \quad \sum_{i=1}^{3} \frac{1}{\tilde{a}_i} = -\frac{\tilde{z}}{\tilde{x}}, \quad \sum_{i=1}^{3} \widetilde{\mathcal{M}}_i = 0, \quad \sum_{i=1}^{3} \tilde{a}_i \widetilde{\mathcal{M}}_i = 0,$$

$$\sum_{i=1}^{3} \tilde{a}_i^2 \widetilde{\mathcal{M}}_i = 1, \quad \sum_{i=1}^{3} \tilde{a}_i^3 \widetilde{\mathcal{M}}_i = 0 \quad \text{and} \quad \sum_{i=1}^{3} \tilde{a}_i^4 \widetilde{\mathcal{M}} = \tilde{z}. \tag{7.82}$$

$$\tilde{\iota} = \tfrac{1}{3}\tilde{a}^3 - \tilde{x}\,\tilde{a}$$

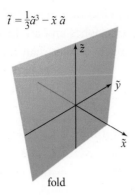

fold

$$\tilde{\iota} = \tfrac{1}{4}\tilde{a}^4 - \tfrac{1}{2}\tilde{z}\,\tilde{a}^2 - \tilde{x}\,\tilde{a}$$

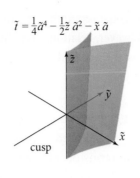

cusp

$$\tilde{\iota} = \tfrac{1}{5}\tilde{a}^5 - \tfrac{1}{3}\tilde{z}\,\tilde{a}^3 - \tfrac{1}{2}\tilde{y}\,\tilde{a}^2 - \tilde{x}\,\tilde{a}$$

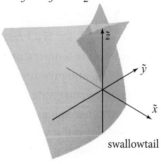

swallowtail

$$\tilde{\iota} = \tfrac{1}{3}\tilde{a}^3 - \tilde{a}\,\tilde{b}^2 - \tilde{z}\,(\tilde{a}^2 + \tilde{b}^2) - \tilde{x}\,\tilde{a} - \tilde{y}\,\tilde{b}$$

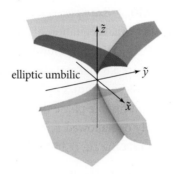

elliptic umbilic

$$\tilde{\iota} = \tfrac{1}{3}(\tilde{a}^3 + \tilde{b}^3) - \tilde{z}\,\tilde{a}\,\tilde{b} - \tilde{x}\,\tilde{a} - \tilde{y}\,\tilde{b}$$

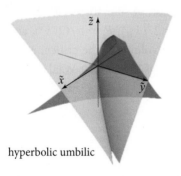

hyperbolic umbilic

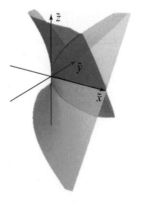

FIGURE 7.15 The five elementary catastrophes (caustic structures) that are possible for a set of light rays specified by one or two state variables $\{\tilde{a}, \tilde{b}\}$ in 3-dimensional space with coordinates (control parameters) $\{\tilde{x}, \tilde{y}, \tilde{z}\}$. The surfaces represent the loci of points of infinite magnification assuming a point source and geometric optics. The actual caustic surfaces will be deformed versions of these basic shapes. The hyperbolic umbilic surfaces are shown from two different viewpoints.

[Hint: You must retain the sign of the magnification.] Of course, not all of these are useful. However, relations like these exist for all catastrophes and are increasingly accurate as the separation of the images becomes much smaller than the scale of variation of $\tilde{t}$.

Exercise 7.15 *Problem: Wavefronts* T2

As we have emphasized, representing light using wavefronts is complementary to treating it in terms of rays. Sketch the evolution of the wavefronts after they propagate through a phase-changing screen and eventually form caustics. Do this for a 2-dimensional cusp, and then consider the formation of a hyperbolic umbilic catastrophe by an astigmatic lens.

Exercise 7.16 ***Example: Van der Waals Catastrophe* T2

The van der Waals equation of state $(P + a/v^2)(v - b) = k_B T$ for H_2O relates the pressure P and specific volume (volume per molecule) v to the temperature T; see Sec. 5.7. Figure 5.8 makes it clear that, at some temperatures T and pressures P, there are three allowed volumes $v(T, P)$, one describing liquid water, one water vapor, and the third an unstable phase that cannot exist in Nature. At other values of T and P, there is only one allowed v. The transition between three allowed v values and one occurs along some curve in the T-P plane—a catastrophe curve.

(a) This curve must correspond to one of the elementary catastrophes explored in the previous exercise. Based on the number of solutions for $v(T, P)$, which catastrophe must it be?

(b) Change variables in the van der Waals equation of state to $p = P/P_c - 1$, $\tau = T/T_c - 1$, and $\rho = v_c/v - 1$, where $T_c = 8a/(27bk_B)$, $P_c = a/(27b^2)$, and $v_c = 3b$ are the temperature, pressure, and specific volume at the critical point C of Fig. 5.8. Show that this change of variables brings the van der Waals equation of state into the form

$$\rho^3 - z\rho - x = 0, \tag{7.83}$$

where $z = -(p/3 + 8\tau/3)$ and $x = 2p/3 - 8\tau/3$.

(c) This equation $\rho^3 - z\rho - x$ is the equilibrium surface associated with the catastrophe-theory potential $t(\rho; x, z) = \frac{1}{4}\rho^4 - \frac{1}{2}z\rho^2 - x\rho$ [Eq. (7.73)]. Correspondingly, the catastrophe [the boundary between three solutions $v(T, P)$ and one] has the universal cusp form $x = \pm 2(z/3)^{2/3}$ [Eq. (7.75)]. Plot this curve in the temperature-pressure plane.

Note that we were guaranteed by catastrophe theory that the catastrophe curve would have this form near its cusp point. However, it is a surprise and quite unusual that, for the van der Waals case, the cusp shape $x = \pm 2(z/3)^{2/3}$ is not confined to the vicinity of the cusp point but remains accurate far from that point.

7.5.2 Aberrations of Optical Instruments [T2]

Much computational effort is expended in the design of expensive optical instruments prior to prototyping and fabrication. This is conventionally discussed in terms of *aberrations,* which provide a perturbative description of rays that complements the singularity-based approach of catastrophe theory. While it is possible to design instruments that take all the rays from a point source S and focus them geometrically onto a point detector D,[11] this is not what is demanded of them in practice. Typically, they have to map an extended image onto an extended surface, for example, a CCD detector. Sometimes the source is large, and the instrument must achieve a large *field of view;* sometimes it is small, and image fidelity close to the axis matters. Sometimes light levels are low, and transmission losses must be minimized. Sometimes the bandwidth of the light is large, and the variation of the imaging with frequency must be minimized. Sometimes diffractive effects are important. The residual imperfections of an instrument are known as *aberrations.*

aberrations

As we have shown, any (geometric-optics) instrument will map, one to many, source points onto detector points. This mapping is usually expanded in terms of a set of basis functions, and several choices are in use, for example, those due to Seidel and Zernike (e.g., Born and Wolf, 1999, Secs. 5.3, 9.2). If we set aside effects caused by the variation of the refractive index with wavelength, known as *chromatic aberration,* there are five common types of geometrical aberration. *Spherical aberration* is the failure to bring a point on the optic axis to a single focus. Instead, an axisymmetric cusp/fold caustic is created. We have already exhibited *astigmatism* in our discussion of the hyperbolic umbilic catastrophe with a non-axisymmetric lens and an axial source (Sec. 7.5.1). It is not hard to make axisymmetric lenses and mirrors, so this does not happen much in practice. However, as soon as we consider off-axis surfaces, we break the symmetry, and astigmatism is unavoidable. *Curvature* arises when the surface on which the rays from point sources are best brought to a focus lies on a curved surface, not on a plane. It is sometimes advantageous to accept this aberration and to curve the detector surface.[12] To understand *coma,* consider a small pencil of rays from an off-axis source that passes through the center of an instrument and is brought to a focus. Now consider a cone of rays about this pencil that passes through the periphery of the lens. When there is coma, these rays will on average be displaced

chromatic aberration

spherical aberration

curvature

coma

---

11. A simple example is to make the interior of a prolate ellipsoidal detector perfectly reflecting and to place S and D at the two foci, as crudely implemented in whispering galleries.

12. For example, in a traditional Schmidt telescope.

distortion

radially. Coma can be ameliorated by reducing the aperture. Finally, there is *distortion*, in which the sides of a square in the source plane are pushed in (*pin cushion*) or out (*barrel*) in the image plane.

7.6 7.6 Gravitational Lenses T2

7.6.1 7.6.1 Gravitational Deflection of Light T2

Albert Einstein's general relativity theory predicts that light rays should be deflected by the gravitational field of the Sun (Ex. 27.3; Sec. 27.2.3). Newton's law of gravity combined with his corpuscular theory of light also predicts this deflection, but through an angle half as great as relativity predicts. A famous measurement, during a 1919 solar eclipse, confirmed the relativistic prediction, thereby making Einstein world famous.

The deflection of light by gravitational fields allows a cosmologically distant galaxy to behave like a crude lens and, in particular, to produce multiple images of a more distant quasar. Many examples of this phenomenon have been observed. The optics of these gravitational lenses provides an excellent illustration of the use of Fermat's principle (e.g., Blandford and Narayan, 1992; Schneider, Ehlers, and Falco, 1992). We explore these issues in this section.

The action of a gravitational lens can only be understood properly using general relativity. However, when the gravitational field is weak, there exists an equivalent Newtonian model, due to Eddington (1919), that is adequate for our purposes. In this model, curved spacetime behaves as if it were spatially flat and endowed with a refractive index given by

refractive index model for gravitational lensing

$$\mathfrak{n} = 1 - \frac{2\Phi}{c^2}, \qquad (7.84)$$

where Φ is the Newtonian gravitational potential, normalized to vanish far from the source of the gravitational field and chosen to have a negative sign (so, e.g., the field at a distance r from a point mass M is $\Phi = -GM/r$). Time is treated in the Newtonian manner in this model. In Sec. 27.2.3, we use a general relativistic version of Fermat's principle to show that for static gravitational fields this index-of-refraction model gives the same predictions as general relativity, up to fractional corrections of order $|\Phi|/c^2$, which are $\lesssim 10^{-5}$ for the lensing examples in this chapter.

second refractive index model

A second Newtonian model gives the same predictions as this index-of-refraction model to within its errors, $\sim |\Phi|/c^2$. We deduce it by rewriting the ray equation (7.48) in terms of Newtonian time t using $ds/dt = C = c/\mathfrak{n}$. The resulting equation is $(\mathfrak{n}^3/c^2)d^2\mathbf{x}/dt^2 = \nabla\mathfrak{n} - 2(\mathfrak{n}/c)^2(d\mathfrak{n}/dt)d\mathbf{x}/dt$. The second term changes the length of the velocity vector $d\mathbf{x}/dt$ by a fractional amount of order $|\Phi|/c^2 \lesssim 10^{-5}$ (so as to keep the length of $d\mathbf{x}/ds$ unity). This is of no significance for our Newtonian model, so we drop this term. The factor $\mathfrak{n}^3 \simeq 1 - 6\Phi/c^2$ produces a fractional correction to $d^2\mathbf{x}/dt^2$ that is of the same magnitude as the fractional errors in our index of refraction model, so we replace this factor by one. The resulting equation of motion for the ray is

$$\frac{d^2\mathbf{x}}{dt^2} = c^2\nabla\mathfrak{n} = -2\nabla\Phi. \tag{7.85}$$

Equation (7.85) says that the photons that travel along rays feel a Newtonian gravitational potential that is twice as large as the potential felt by low-speed particles; the photons, moving at speed c (aside from fractional changes of order $|\Phi|/c^2$), respond to that doubled Newtonian field in the same way as any Newtonian particle would. The extra deflection is attributable to the geometry of the spatial part of the metric being non-Euclidean (Sec. 27.2.3).

7.6.2 Optical Configuration T2

To understand how gravitational lenses work, we adopt some key features from our discussion of optical catastrophes formed by an intervening screen. However, there are some essential differences.

- The source is not assumed to be distant from the screen (which we now call a lens, L).

- Instead of tracing rays emanating from a point source S, we consider a *congruence* of rays emanating from the observer O (i.e., us) and propagating backward in time past the lens to the sources. This is because there are many stars and galaxies whose images will be distorted by the lens. The caustics envelop the sources.

- The universe is expanding, which makes the optics formally time-dependent. However, as we discuss in Sec. 28.6.2, we can work in comoving coordinates and still use Fermat's principle. For the moment, we introduce three distances: d_{OL} for distance from the observer to the lens, d_{OS} for the distance from the observer to the source, and d_{LS} for the distance from the lens to the source. We evaluate these quantities cosmologically in Sec. 28.6.2.

- Instead of treating a and b as the state variables that describe rays (Sec. 7.5.1), we use a 2-dimensional (small) angular vector $\boldsymbol{\theta}$ measuring the image position on the sky. We also replace the control parameters x and y with the 2-dimensional angle $\boldsymbol{\beta}$, which measures the location that the image of the source would have in the absence of the lens. We can also treat the distance d_{OS} as a third control parameter replacing z.

The Hessian matrix, replacing Eq. (7.80), is now the Jacobian of the vectorial angles that a small, finite source would subtend in the absence and in the presence of the lens:

$$\mathcal{H} = \mathcal{M}^{-1} = \frac{\partial\boldsymbol{\beta}}{\partial\boldsymbol{\theta}}. \tag{7.86}$$

As the specific intensity $I_\nu = dE/dA\,dt\,d\nu\,d\Omega$ is conserved along a ray (see Sec. 3.6), the determinant of $\mathcal{H}$ is just the ratio of the flux of energy per unit frequency without the lens to the flux with the lens, and correspondingly the determinant of $\mathcal{M}$ (the scalar magnification) is the ratio of flux with the lens to that without the lens.

7.6.3

7.6.3 Microlensing T2

Our first example of a gravitational lens is a point mass—specifically, a star. This phenomenon is known as *microlensing*, because the angles of deflection are typically microarcseconds.[13] The source is also usually another star, which we also treat as a point.

We first compute the deflection of a Newtonian particle with speed v passing by a mass M with impact parameter b. By computing the perpendicular impulse, it is straightforward to show that the deflection angle is $2GM/v^2$. Replacing v by c and doubling the answer gives the small deflection angle for light:

microlensing deflection angle

$$\boldsymbol{\alpha} = \frac{4GM}{bc^2} = 1.75\left(\frac{M}{M_\odot}\right)\left(\frac{b}{R_\odot}\right)^{-1}\hat{\mathbf{b}} \text{ arcsec,} \tag{7.87}$$

where $\hat{\mathbf{b}}$ is a unit vector along the impact parameter, which allows us to treat the deflection as a 2-dimensional vector like $\boldsymbol{\theta}$ and $\boldsymbol{\beta}$. $M_\odot$ and $R_\odot$ are the solar mass and radius, respectively.

microlensing lens equation

The imaging geometry can be expressed as a simple vector equation called the *lens equation* (Fig. 7.16):

$$\boldsymbol{\theta} = \boldsymbol{\beta} + \frac{d_{LS}}{d_{OS}}\boldsymbol{\alpha}. \tag{7.88}$$

A point mass exhibits circular symmetry, so we can treat this equation as a scalar equation and rewrite it in the form

$$\theta = \beta + \frac{\theta_E^2}{\theta}, \tag{7.89}$$

where

$$\theta_E = \left(\frac{4GM}{d_{\text{eff}}c^2}\right)^{1/2} = 903\left(\frac{M}{M_\odot}\right)^{1/2}\left(\frac{d_{\text{eff}}}{10\text{ kpc}}\right)^{-1/2}\mu \text{ arcsec} \tag{7.90}$$

is the *Einstein radius*, and

$$d_{\text{eff}} = \frac{d_{OL}d_{OS}}{d_{LS}} \tag{7.91}$$

is the *effective distance*. (Here 10 kpc means 10 kiloparsecs, about 30,000 light years.)

---

13. Interestingly, Newton speculated that light rays could be deflected by gravity, and the underlying theory of microlensing was worked out correctly by Einstein in 1912, before he realized that the deflection was twice the Newtonian value.

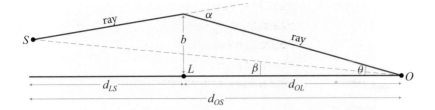

FIGURE 7.16 Geometry for microlensing of a stellar source S by a stellar lens L observed at O.

The solutions to this quadratic equation are

$$\theta_\pm = \frac{\beta}{2} \pm \sqrt{\theta_E^2 + \left(\frac{\beta}{2}\right)^2}. \tag{7.92}$$

image locations

The magnification of the two images can be computed directly by evaluating the reciprocal of the determinant of $\mathcal{H}$ from Eq. (7.86). However, it is quicker to exploit the circular symmetry and note that the element of source solid angle in polar coordinates is $\beta d\beta d\phi$, while the element of image solid angle is $\theta d\theta d\phi$, so that

$$\mathcal{M} = \frac{\theta \, d\theta}{\beta \, d\beta} = \frac{1}{1 - (\theta_E/\theta)^4}. \tag{7.93}$$

image magnifications

The eigenvalues of the magnification matrix are $[1 + (\theta_E/\theta)^2]^{-1}$ and $[1 - (\theta_E/\theta)^2]^{-1}$. The former describes the radial magnification, the latter the tangential magnification. As $\beta \to 0, \theta \to \theta_E$ for both images. If we consider a source of finite angular size, when β approaches this size then the tangential stretching is pronounced and two nearly circular arcs are formed on opposite sides of the lens, one with $\theta > \theta_E$; the other, inverted, with $\theta < \theta_E$. When β is reduced even more, the arcs join up to form an *Einstein ring* (Fig. 7.17).

Einstein ring

Astronomers routinely observe microlensing events when stellar sources pass behind stellar lenses. They are unable to distinguish the two images and so measure a combined magnification $\mathcal{M} = |\mathcal{M}_+| + |\mathcal{M}_-|$. If we substitute β for θ, then we have

$$\mathcal{M} = \frac{(\theta_E^2 + \frac{1}{2}\beta^2)}{(\theta_E^2 + \frac{1}{4}\beta^2)^{1/2}\beta}. \tag{7.94}$$

Note that in the limit of large magnifications, $\mathcal{M} \sim \theta_E/\beta$, and the probability that the magnification exceeds $\mathcal{M}$ is proportional to the cross sectional area $\pi\beta^2 \propto \mathcal{M}^{-2}$ (cf. Fig. 7.16).

If the speed of the source relative to the continuation of the O-L line is v, and the closest approach to this line is h, which happens at time $t = 0$, then $\beta = (h^2 + v^2t^2)^{1/2}/d_{OS}$, and then there is a one-parameter family of magnification curves (shown in Fig. 7.18). The characteristic variation of the magnification can be used to distinguish this phenomenon from intrinsic stellar variation. This behavior is very

FIGURE 7.17 The source LRG 3-757, as imaged by the Hubble Space Telescope. The blue Einstein ring is the image of two background galaxies formed by the gravitational field associated with the intervening central (yellow) lens galaxy. The accurate alignment of the lens and source galaxies is quite unusual. (ESA/Hubble and NASA.)

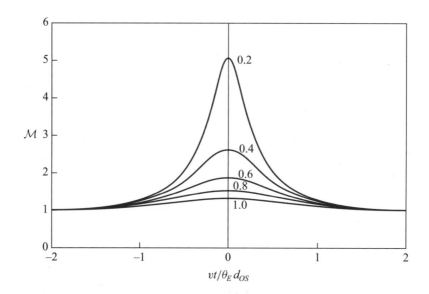

FIGURE 7.18 Variation of the combined magnifications of a stellar source as it passes behind a lens star. The time t is measured in units of $d_{OS}\theta_E/v$, and the parameter that labels the curves is $h/(d_{OS}\theta_E)$.

different from that at a generic caustic (Sec. 7.5.1) and is not structurally stable: if the axisymmetry is broken, then the behavior will change significantly. Nevertheless, for finite-sized sources of light and stars or nearly circular galaxies, stable Einstein rings are commonly formed.

Exercise 7.17 *Example: Microlensing Time Delay* T2

An alternative derivation of the lens equation for a point-mass lens, Eq. (7.88), evaluates the time delay along a path from the source to the observer and finds the true ray by extremizing it with respect to variations of θ [cf. Eq. (7.68)].

(a) Show that the geometric time delay is given by

$$t_{\text{geo}} = \frac{1}{2c} d_{\text{eff}} (\boldsymbol{\theta} - \boldsymbol{\beta})^2. \tag{7.95}$$

(b) Next show that the lens time delay can be expressed as

$$t_{\text{lens}} = -(4GM/c^3) \ln b + \text{const},$$

where b is the impact parameter. (It will be helpful to evaluate the difference in delays between two rays with differing impact parameters.) This is known as the *Shapiro delay* and is discussed further in Sec. 27.2.4.

(c) Show that the lens delay can also be written as

$$t_{\text{lens}} = -\frac{2}{c^3} \int dz \Phi = -\frac{2}{c^3} \Phi_2, \tag{7.96}$$

where Φ_2 is the surface gravitational potential obtained by integrating the 3-dimensional potential Φ along the path. The surface potential is only determined up to an unimportant, divergent constant, which is acceptable because we are only interested in dt_{lens}/db which is finite.

(d) By minimizing $t_{\text{geo}} + t_{\text{lens}}$, derive the lens equation (7.88).

Exercise 7.18 *Derivation: Microlensing Variation* T2
Derive Eq. (7.94).

Exercise 7.19 *Problem: Magnification by a Massive Black Hole* T2
Suppose that a large black hole forms two images of a background source separated by an angle θ. Let the fluxes of the two images be F_+ and $F_- < F_+$. Show that the flux from the source would be $F_+ - F_-$ if there were no lens and that the black hole should be located an angular distance $[1 + (F_-/F_+)^{-1/2}]^{-1}\theta$ along the line from the brighter image to the fainter one. (Only consider small angle deflections.)

7.6.4 Lensing by Galaxies T2

Most observed gravitational lenses are galaxies. Observing these systems brings out new features of the optics and proves useful for learning about galaxies and the universe. Galaxies comprise dark matter and stars, and the dispersion $\langle v_{\parallel}^2 \rangle$ in the stars' velocities along the line of sight can be measured using spectroscopy. The virial theorem (Goldstein, Poole, and Safko, 2002) tells us that the kinetic energy of the matter in the galaxy is half the magnitude of its gravitational potential energy Φ. We can therefore make an order of magnitude estimate of the ray deflection angle caused by a galaxy by using Eq. (7.87):

$$\alpha \sim \frac{4GM}{bc^2} \sim \frac{4|\Phi|}{c^2} \sim 4 \times 2 \times \frac{3}{2} \times \frac{\langle v_{\parallel}^2 \rangle}{c^2} \sim \frac{12 \langle v_{\parallel}^2 \rangle}{c^2}. \tag{7.97}$$

This evaluates to $\alpha \sim 2$ arcsec for a typical galaxy velocity dispersion of ~ 300 km s^{-1}. The images can typically be resolved using radio and optical telescopes, but their separations are much less than the full angular sizes of distant galaxies and so the imaging is sensitive to the lens galaxy's structure. As the lens is typically far more complicated than a point mass, it is now convenient to measure the angle θ relative to the reference ray that connects us to the source.

We can describe the optics of a galaxy gravitational lens by adapting the formalism that we developed for a point-mass lens in Ex. 7.17. We assume that there is a point source at β and consider paths designated by θ. The geometrical time delay is unchanged. In Ex. 7.17, we showed that the lens time delay for a point mass was proportional to the surface gravitational potential Φ_2. A distributed lens is handled by adding the potentials associated with all the point masses out of which it can be considered as being composed. In other words, we simply use the surface potential for the distributed mass in the galaxy,

$$t = t_{\text{geo}} + t_{\text{lens}} = \frac{d_{\text{eff}}}{2c}(\theta - \beta)^2 - \frac{2}{c^3}\Phi_2 = \frac{d_{\text{eff}}\tilde{t}}{c}, \tag{7.98}$$

scaled time delay for lensing by galaxies

where the *scaled time delay* t is defined by $\tilde{t}(\theta; \beta) = \frac{1}{2}(\theta - \beta)^2 - \Psi(\theta)$, and $\Psi = 2\Phi_2/(c^2 d_{\text{eff}})$. The quantity Ψ satisfies the 2-dimensional Poisson equation:

$$\nabla_{2,\theta}^2 \Psi = \frac{8\pi G \Sigma}{d_{\text{eff}}c^2}, \tag{7.99}$$

where Σ is the density of matter per unit solid angle, and the 2-dimensional laplacian describes differentiation with respect to the components of θ.[14]

As written, Eq. (7.98) describes all paths for a given source position, only a small number of which correspond to true rays. However, if, instead, we set $\beta = 0$ and choose any convenient origin for θ so that

---

14. The minimum surface density (expressed as mass per area) of a cosmologically distant lens needed to produce multiple images of a background source turns out to be ~ 1 g cm^{-2}. It is remarkable that such a seemingly small surface density operating on these scales can make such a large difference to our view of the universe. It is tempting to call this a "rule of thumb," because it is roughly the column density associated with one's thumb!

$$\tilde{t}(\boldsymbol{\theta}) = \frac{1}{2}\theta^2 - \Psi(\boldsymbol{\theta}),\qquad(7.100)$$

then three useful features of $\tilde{t}$ emerge:

- if there is an image at $\boldsymbol{\theta}$, then the source position is simply $\boldsymbol{\beta} = \nabla_\theta \tilde{t}$ [cf. Eq. (7.88)];

- the magnification tensor associated with this image can be calculated by taking the inverse of the Hessian matrix $\mathcal{H} = \nabla_\theta \nabla_\theta \tilde{t}$ [cf. Eq. (7.86)]; and

- the measured differences in the times of variation observed in multiple images of the same source are just the differences in $d_{\mathrm{eff}}\tilde{t}/c$ evaluated at the image positions.[15]

Computing $\tilde{t}$ for a model of a putative lens galaxy allows one to assess whether background sources are being multiply imaged and, if so, to learn about the lens as well as the source.

EXERCISES

Exercise 7.20 *Problem: Catastrophe Optics of an Elliptical Gravitational Lens* T2
Consider an elliptical gravitational lens where the potential Ψ is modeled by

$$\Psi(\boldsymbol{\theta}) = (1 + A\theta_x^2 + 2B\theta_x\theta_y + C\theta_y^2)^q; \quad 0 < q < 1/2.\qquad(7.101)$$

Determine the generic form of the caustic surfaces, the types of catastrophe encountered, and the change in the number of images formed when a point source crosses these surfaces. Note that it is in the spirit of catastrophe theory *not* to compute exact expressions but to determine scaling laws and to understand the qualitative features of the images.

Exercise 7.21 *Challenge: Microlensing in a Galaxy* T2
Our discussion of microlensing assumed a single star and a circularly symmetric potential about it. This is usually a good approximation for stars in our galaxy. However, when the star is in another galaxy and the source is a background quasar (Figs. 7.19, 7.20), it is necessary to include the gravitational effects of the galaxy's other stars and its dark matter. Recast the microlensing analysis (Sec. 7.6.3) in the potential formulation of Eq. (7.98) and add *external magnification* and *external shear* contributions to

---

15. These differences can be used to measure the size and age of the universe. To order of magnitude, the relative delays are the times it takes light to cross the universe (or equivalently, the age of the universe, roughly 10 Gyr) times the square of the scattering angle (roughly 2 arcsec or $\sim 10^{-5}$ radians), which is roughly 1 year. This is very convenient for astronomers (see Fig. 7.20).

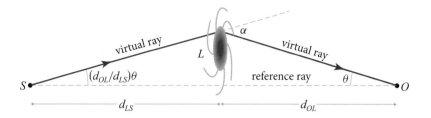

FIGURE 7.19 Geometry for gravitational lensing of a quasar source S by a galaxy lens L observed at O.

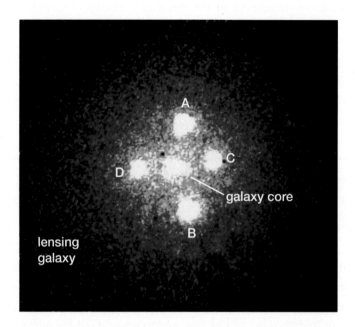

FIGURE 7.20 Gravitational lens in which a distant quasar, Q2237+0305, is quadruply imaged by an intervening galaxy. The four quasar images are denoted A, B, C, and D. The galaxy is much larger than the separation of the images (1–1.5 arcsec) but its bright central core is labeled. There is also a fifth, faint image coincident with this core. There are many examples of gravitational lenses like this, where the source region is very compact and variable so that the delay in the variation seen in the individual images can be used to measure the distance of the source and the size and age of the universe. In addition, microlensing-like variations in the images are induced by individual stars in the lens galaxy moving in front of the quasar. Analyzing these changes can be used to measure the proportion of stars and dark matter in galaxies. Adapted from image by Hubble Space Telescope. (NASA, ESA, STScI.)

Ψ that are proportional to $\theta_x^2 + \theta_y^2$ and $\theta_x^2 - \theta_y^2$, respectively. The latter will break the circular symmetry, and structurally stable caustics will be formed. Explore the behavior of these caustics as you vary the strength and sign of the magnification and shear contributions. Plot a few flux variations that might be observed.

7.7 Polarization

In our geometric-optics analyses thus far, we have either dealt with a scalar wave (e.g., a sound wave) or simply supposed that individual components of vector or tensor waves can be treated as scalars. For most purposes, this is indeed the case, and we continue to use this simplification in the following chapters. However, there are some important wave properties that are unique to vector (or tensor) waves. Most of these come under the heading of *polarization* effects. In Secs. 27.3, 27.4, and 27.5, we study polarization effects for (tensorial) gravitational waves. Here and in Secs. 10.5 and 28.6.1, we examine them for electromagnetic waves.

An electromagnetic wave's two polarizations are powerful tools for technology, engineering, and experimental physics. However, we forgo any discussion of this in the present chapter. Instead we focus solely on the geometric-optics propagation law for polarization (Sec. 7.7.1) and an intriguing aspect of it—the geometric phase (Sec. 7.7.2).

7.7.1 Polarization Vector and Its Geometric-Optics Propagation Law

A plane electromagnetic wave in a vacuum has its electric and magnetic fields $\mathbf{E}$ and $\mathbf{B}$ perpendicular to its propagation direction $\hat{\mathbf{k}}$ and perpendicular to each other. In a medium, $\mathbf{E}$ and $\mathbf{B}$ may or may not remain perpendicular to $\hat{\mathbf{k}}$, depending on the medium's properties. For example, an Alfvén wave has its vibrating magnetic field perpendicular to the background magnetic field, which can make an arbitrary angle with respect to $\hat{\mathbf{k}}$. By contrast, in the simplest case of an isotropic dielectric medium, where the dispersion relation has our standard dispersion-free form $\Omega = (c/\mathfrak{n})k$, the group and phase velocities are parallel to $\hat{\mathbf{k}}$, and $\mathbf{E}$ and $\mathbf{B}$ turn out to be perpendicular to $\hat{\mathbf{k}}$ and to each other—as in a vacuum. In this section, we confine attention to this simple situation and to linearly polarized waves, for which $\mathbf{E}$ oscillates linearly along a unit polarization vector $\hat{\mathbf{f}}$ that is perpendicular to $\hat{\mathbf{k}}$:

polarization vector

$$\mathbf{E} = A e^{i\varphi}\,\hat{\mathbf{f}}, \qquad \hat{\mathbf{f}} \cdot \hat{\mathbf{k}} \equiv \hat{\mathbf{f}} \cdot \nabla\varphi = 0. \tag{7.102}$$

In the eikonal approximation, $A e^{i\varphi} \equiv \psi$ satisfies the geometric-optics propagation laws of Sec. 7.3, and the polarization vector $\hat{\mathbf{f}}$, like the amplitude A, will propagate along the rays. The propagation law for $\hat{\mathbf{f}}$ can be derived by applying the eikonal approximation to Maxwell's equations, but it is easier to infer that law by simple physical reasoning:

1. Since $\hat{\mathbf{f}}$ is orthogonal to $\hat{\mathbf{k}}$ for a plane wave, it must also be orthogonal to $\hat{\mathbf{k}}$ in the eikonal approximation (which, after all, treats the wave as planar on lengthscales long compared to the wavelength).

2. If the ray is straight, then the medium, being isotropic, is unable to distinguish a slow right-handed rotation of $\hat{\mathbf{f}}$ from a slow left-handed rotation, so there will be no rotation at all: $\hat{\mathbf{f}}$ will continue always to point in the same direction (i.e., $\hat{\mathbf{f}}$ will be kept parallel to itself during transport along the ray).

3. If the ray bends, so $d\hat{\mathbf{k}}/ds \neq 0$ (where s is distance along the ray), then $\hat{\mathbf{f}}$ will have to change as well, so as always to remain perpendicular to $\hat{\mathbf{k}}$. The direction of $\hat{\mathbf{f}}$'s change must be $\hat{\mathbf{k}}$, since the medium, being isotropic, cannot provide any other preferred direction for the change. The magnitude of the change is determined by the requirement that $\hat{\mathbf{f}} \cdot \hat{\mathbf{k}}$ remain zero all along the ray and that $\hat{\mathbf{k}} \cdot \hat{\mathbf{k}} = 1$. This immediately implies that the propagation law for $\hat{\mathbf{f}}$ is

$$\frac{d\hat{\mathbf{f}}}{ds} = -\hat{\mathbf{k}}\left(\hat{\mathbf{f}} \cdot \frac{d\hat{\mathbf{k}}}{ds}\right).$$

(7.103)

propagation law for polarization vector

This equation states that the vector $\hat{\mathbf{f}}$ is parallel-transported along the ray (cf. Fig. 7.5 in Sec. 24.3.3). Here "parallel transport" means: (i) Carry $\hat{\mathbf{f}}$ a short distance along the ray, keeping it parallel to itself in 3-dimensional space. Because of the bending of the ray and its tangent vector $\hat{\mathbf{k}}$, this will cause $\hat{\mathbf{f}}$ to no longer be perpendicular to $\hat{\mathbf{k}}$. (ii) Project $\hat{\mathbf{f}}$ perpendicular to $\hat{\mathbf{k}}$ by adding onto it the appropriate multiple of $\hat{\mathbf{k}}$. (The techniques of differential geometry for curved lines and surfaces, which we develop in Chaps. 24 and 25 in preparation for studying general relativity, give powerful mathematical tools for analyzing this parallel transport.)

7.7.2

7.7.2 Geometric Phase T2

We use the polarization propagation law (7.103) to illustrate a quite general phenomenon known as the *geometric phase,* or sometimes as the *Berry phase,* after Michael Berry who elucidated it. For further details and some history of this concept, see Berry (1990).

As a simple context for the geometric phase, consider a linearly polarized, monochromatic light beam that propagates in an optical fiber. Focus on the evolution of the polarization vector along the fiber's optic axis. We can imagine bending the fiber into any desired shape, thereby controlling the shape of the ray. The ray's shape in turn will control the propagation of the polarization via Eq. (7.103).

If the fiber and ray are straight, then the propagation law (7.103) keeps $\hat{\mathbf{f}}$ constant. If the fiber and ray are circular, then Eq. (7.103) causes $\hat{\mathbf{f}}$ to rotate in such a way as to always point along the generator of a cone, as shown in Fig. 7.21a. This polarization behavior, and that for any other ray shape, can be deduced with the aid of a unit sphere on which we plot the ray direction $\hat{\mathbf{k}}$ (Fig. 7.21b). For example, the ray directions at ray locations C and H of panel a are as shown in panel b of the figure. Notice that the trajectory of $\hat{\mathbf{k}}$ around the unit sphere is a great circle.

On the unit sphere we also plot the polarization vector $\hat{\mathbf{f}}$—one vector at each point corresponding to a ray direction. Because $\hat{\mathbf{f}} \cdot \hat{\mathbf{k}} = 0$, the polarization vectors are always tangent to the unit sphere. Notice that each $\hat{\mathbf{f}}$ on the unit sphere is identical in length and direction to the corresponding one in the physical space of Fig. 7.21a.

The parallel-transport law (7.103) keeps constant the angle α between $\hat{\mathbf{f}}$ and the trajectory of $\hat{\mathbf{k}}$ (i.e., the great circle in panel b of the figure). Translated back to

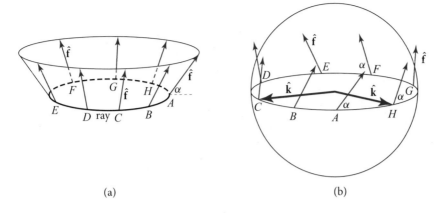

(a)   (b)

FIGURE 7.21 (a) The ray along the optic axis of a circular loop of optical fiber, and the polarization vector $\hat{\mathbf{f}}$ that is transported along the ray by the geometric-optics transport law $d\hat{\mathbf{f}}/ds = -\hat{\mathbf{k}}(\hat{\mathbf{f}} \cdot d\hat{\mathbf{k}}/ds)$. (b) The polarization vector $\hat{\mathbf{f}}$ drawn on the unit sphere. The vector from the center of the sphere to each of the points $A, B, \ldots, H$ is the ray's propagation direction $\hat{\mathbf{k}}$, and the polarization vector (which is orthogonal to $\hat{\mathbf{k}}$ and thus tangent to the sphere) is identical to that in the physical space of the ray (panel a).

panel a, this constancy of α implies that the polarization vector points always along the generators of the cone, whose opening angle is $\pi/2 - \alpha$, as shown.

Next let the fiber and its central axis (the ray) be helical as shown in Fig. 7.22a. In this case, the propagation direction $\hat{\mathbf{k}}$ rotates, always maintaining the same angle θ to the vertical direction, and correspondingly its trajectory on the unit sphere of Fig. 7.22b is a circle of constant polar angle θ. Therefore (as one can see, e.g., with the aid of a large globe of Earth and a pencil transported around a circle of latitude $90° - \theta$), the parallel-transport law dictates that the angle α between $\hat{\mathbf{f}}$ and the circle *not* remain constant, but instead rotate at the rate

$$d\alpha/d\phi = \cos\theta. \tag{7.104}$$

Here ϕ is the angle (longitude on the globe) around the circle. This is the same propagation law as for the direction of swing of a Foucault pendulum as Earth turns (cf. Box 14.5), and for the same reason: the gyroscopic action of the Foucault pendulum is described by parallel transport of its plane along Earth's spherical surface.

In the case where θ is arbitrarily small (a nearly straight ray), Eq. (7.104) says $d\alpha/d\phi = 1$. This is easily understood: although $\hat{\mathbf{f}}$ remains arbitrarily close to constant, the trajectory of $\hat{\mathbf{k}}$ turns rapidly around a tiny circle about the pole of the unit sphere, so α changes rapidly—by a total amount $\Delta\alpha = 2\pi$ after one trip around the pole, $\Delta\phi = 2\pi$; whence $d\alpha/d\phi = \Delta\alpha/\Delta\phi = 1$. For any other helical pitch angle θ, Eq. (7.104) says that during one round trip, α will change by an amount $2\pi\cos\theta$ that lags behind its change for a tiny circle (nearly straight ray) by the lag angle $\alpha_{\text{lag}} = 2\pi(1 - \cos\theta)$, which is also the solid angle $\Delta\Omega$ enclosed by the path of $\hat{\mathbf{k}}$ on the unit sphere:

$$\alpha_{\text{lag}} = \Delta\Omega. \tag{7.105}$$

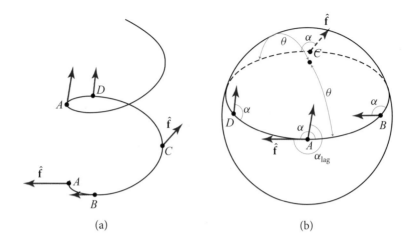

FIGURE 7.22 (a) The ray along the optic axis of a helical loop of optical fiber, and the polarization vector $\hat{\mathbf{f}}$ that is transported along this ray by the geometric-optics transport law $d\hat{\mathbf{f}}/ds = -\hat{\mathbf{k}}(\hat{\mathbf{f}} \cdot d\hat{\mathbf{k}}/ds)$. The ray's propagation direction $\hat{\mathbf{k}}$ makes an angle $\theta = 73°$ to the vertical direction. (b) The trajectory of $\hat{\mathbf{k}}$ on the unit sphere (a circle with polar angle $\theta = 73°$), and the polarization vector $\hat{\mathbf{f}}$ that is parallel transported along that trajectory. The polarization vectors in panel a are deduced from the parallel-transport law demonstrated in panel b. The lag angle $\alpha_{\text{lag}} = 2\pi(1 - \cos\theta) = 1.42\pi$ is equal to the solid angle contained inside the trajectory of $\hat{\mathbf{k}}$ (the $\theta = 73°$ circle).

(For the circular ray of Fig. 7.21, the enclosed solid angle is $\Delta\Omega = 2\pi$ steradians, so the lag angle is 2π radians, which means that $\hat{\mathbf{f}}$ returns to its original value after one trip around the optical fiber, in accord with the drawings in the figure.)

lag angle for polarization vector in an optical fiber

Remarkably, Eq. (7.105) is true for light propagation along an optical fiber of any shape: if the light travels from one point on the fiber to another at which the tangent vector $\hat{\mathbf{k}}$ has returned to its original value, then the lag angle is given by the enclosed solid angle on the unit sphere, Eq. (7.105).

By itself, the relationship $\alpha_{\text{lag}} = \Delta\Omega$ is merely a cute phenomenon. However, it turns out to be just one example of a very general property of both classical and quantum mechanical systems when they are forced to make slow, *adiabatic* changes described by circuits in the space of parameters that characterize them. In the more general case, one focuses on a phase lag rather than a direction-angle lag. We can easily translate our example into such a phase lag.

geometric phase change

The apparent rotation of $\hat{\mathbf{f}}$ by the lag angle $\alpha_{\text{lag}} = \Delta\Omega$ can be regarded as an advance of the phase of one circularly polarized component of the wave by $\Delta\Omega$ and a phase retardation of the other circular polarization by the same amount. Thus the phase of a circularly polarized wave will change, after one circuit around the fiber's helix, by an amount equal to the usual phase advance $\Delta\varphi = \int \mathbf{k} \cdot d\mathbf{x}$ (where $d\mathbf{x}$ is displacement along the fiber) plus an extra *geometric* phase change $\pm\Delta\Omega$, where the sign is given by the sense of circular polarization. This type of geometric phase change is found quite generally, when classical vector or tensor waves propagate through backgrounds that change slowly, either temporally or spatially. The phases of the wave functions of quantum mechanical particles with spin behave similarly.

Exercise 7.22 *Derivation: Parallel Transport* T2
Use the parallel-transport law (7.103) to derive the relation (7.104).

Exercise 7.23 *Problem: Martian Rover* T2
A Martian Rover is equipped with a single gyroscope that is free to pivot about the direction perpendicular to the plane containing its wheels. To climb a steep hill on Mars without straining its motor, it must circle the summit in a decreasing spiral trajectory. Explain why there will be an error in its measurement of North after it has reached the summit. Could it be programmed to navigate correctly? Will a stochastic error build up as it traverses a rocky terrain?

Bibliographic Note

Modern textbooks on optics deal with the geometric-optics approximation only for electromagnetic waves propagating through a dispersion-free medium. Accordingly, they typically begin with Fermat's principle and then treat in considerable detail the paraxial approximation, applications to optical instruments, and sometimes the human eye. There is rarely any mention of the eikonal approximation or of multiple images and caustics. Examples of texts of this sort that we like are Bennett (2008), Ghatak (2010), and Hecht (2017). For a far more thorough treatise on geometric optics of scalar and electromagnetic waves in isotropic and anisotropic dielectric media, see Kravtsov (2005). A good engineering-oriented text with many contemporary applications is Iizuka (1987).

We do not know of textbooks that treat the eikonal approximation to the degree of generality used in this chapter, though some should, since it has applications to all types of waves (many of which are explored later in this book). For the eikonal approximation specialized to Maxwell's equations, see Kravtsov (2005) and the classic treatise on optics by Born and Wolf (1999), which in this new edition has modern updates by a number of other authors. For the eikonal approximation specialized to the Schrödinger equation and its connection to Hamilton-Jacobi theory, see most any quantum mechanics textbook (e.g., Griffiths, 2004).

Multiple-image formation and caustics are omitted from most standard optics textbooks, except for a nice but out-of-date treatment in Born and Wolf (1999). Much better are the beautiful review by Berry and Upstill (1980) and the much more thorough treatments in Kravtsov (2005) and Nye (1999). For an elementary mathematical treatment of catastrophe theory, we like Saunders (1980). For a pedagogical treatise on gravitational lenses, see Schneider, Ehlers, and Falco (1992). Finally, for some history and details of the geometric phase, see Berry (1990).

CHAPTER EIGHT

Diffraction

I have seen—without any illusion—three broad stripes in the spectrum of Sirius,
which seem to have no similarity to those of sunlight.

JOSEPH VON FRAUNHOFER (1814–1815)

8.1 Overview

The previous chapter was devoted to the classical mechanics of wave propagation. We showed how a classical wave equation can be solved in the short-wavelength (eikonal) approximation to yield Hamilton's dynamical equations for its rays. When the medium is time independent (as we require in this chapter), we showed that the frequency of a wave packet is constant, and we imported a result from classical mechanics—the principle of stationary action—to show that the true geometric-optics rays coincide with paths along which the action (the phase) is stationary [Eqs. (7.44) and (7.45a) and associated discussion]. Our physical interpretation of this result was that the waves do indeed travel along every path, from some source to a point of observation, where they are added together, but they only give a significant net contribution when they can add coherently in phase—along the true rays [Eq. (7.45b)]. Essentially, this is Huygens' model of wave propagation, or, in modern language, a *path integral*.

Huygens' principle asserts that every point on a wavefront acts as a source of secondary waves that combine so their envelope constitutes the advancing wavefront. This principle must be supplemented by two ancillary conditions: that the secondary waves are only formed in the forward direction, not backward, and that a $\pi/2$ phase shift be introduced into the secondary wave. The reason for the former condition is obvious, that for the latter, less so. We discuss both together with the formal justification of Huygens' construction in Sec. 8.2.

We begin our exploration of the "wave mechanics" of optics in this chapter and continue it in Chaps. 9 and 10. Wave mechanics differs increasingly from geometric optics as the reduced wavelength λ increases relative to the lengthscales $\mathcal{R}$ of the phase fronts and $\mathcal{L}$ of the medium's inhomogeneities. The number of paths that can combine constructively increases, and the rays that connect two points become blurred. In quantum mechanics, we recognize this phenomenon as the uncertainty principle, and it is just as applicable to photons as to electrons.

- This chapter depends substantially on Secs. 7.1–7.4 (geometric optics).

- In addition, Sec. 8.6 of this chapter (on diffraction at a caustic) depends on Sec. 7.5.

- Chapters 9 and 10 depend substantially on Secs. 8.1–8.5 of this chapter.

- Nothing else in this book relies on this chapter.

Solving the wave equation exactly is very hard, except in simple circumstances. Geometric optics is one approximate method of solving it—a method that works well in the short-wavelength limit. In this chapter and the next two, we develop approximate techniques that work when the wavelength becomes longer and geometric optics fails.

In this book, we make a somewhat artificial distinction between phenomena that arise when an effectively infinite number of propagation paths are involved (which we call *diffraction* and describe in this chapter) and those when a few paths, or, more correctly, a few tight bundles of rays are combined (which we term *interference,* and whose discussion we defer to the next chapter).

In Sec. 8.2, we present the Fresnel-Helmholtz-Kirchhoff theory that underlies most elementary discussions of diffraction. We then distinguish between Fraunhofer diffraction (the limiting case when spreading of the wavefront mandated by the uncertainty principle is important) and Fresnel diffraction (where wavefront spreading is a modest effect, and geometric optics is beginning to work, at least roughly). In Sec. 8.3, we illustrate Fraunhofer diffraction by computing the angular resolution of the Hubble Space Telescope, and in Sec. 8.4, we analyze Fresnel diffraction and illustrate it using zone plates and lunar occultation of radio waves.

Many contemporary optical devices can be regarded as linear systems that take an input wave signal and transform it into a linearly related output. Their operation, particularly as image-processing devices, can be considerably enhanced by processing the signal in the spatial Fourier domain, a procedure known as spatial filtering. In Sec. 8.5 we introduce a tool for analyzing such devices: paraxial Fourier optics— a close analog of the paraxial geometric optics of Sec. 7.4. Using paraxial Fourier optics, we develop the theory of image processing by spatial filters and use it to discuss various types of filters and the phase-contrast microscope. We also use Fourier optics to develop the theory of Gaussian beams—the kind of light beam produced by lasers when (as is usual) their optically resonating cavities have spherical mirrors. Finally, in Sec. 8.6 we analyze diffraction near a caustic of a wave's phase field, a location where geometric optics predicts a divergent magnification of the wave (Sec. 7.5). As we shall see, diffraction keeps the magnification finite and produces an oscillating energy-flux pattern (interference fringes).

8.2 Helmholtz-Kirchhoff Integral

In this section, we derive a formalism for describing diffraction. We restrict attention to the simplest (and, fortunately, the most widely useful) case: a monochromatic scalar wave,

$$\Psi = \psi(\mathbf{x})e^{-i\omega t}, \tag{8.1a}$$

with field variable ψ that satisfies the Helmholtz equation

$$\nabla^2\psi + k^2\psi = 0, \quad \text{with } k = \omega/c, \tag{8.1b}$$

Helmholtz equation

except at boundaries. Generally, Ψ will represent a real-valued physical quantity, but for mathematical convenience we give it a complex representation and take the real part of Ψ when making contact with physical measurements. This is in contrast to a quantum mechanical wave function satisfying the Schrödinger equation, which is an intrinsically complex function. We assume that the wave in Eqs. (8.1) is monochromatic (constant ω) and nondispersive, and that the medium is isotropic and homogeneous (phase speed equal to group speed, and both with a constant value C, so k is also constant). Each of these assumptions can be relaxed, but with some technical penalty.

The scalar formalism that we develop based on Eq. (8.1b) is fully valid for weak sound waves in a homogeneous fluid or solid (e.g., air; Secs. 12.2 and 16.5). It is also quite accurate, but not precisely so, for the most widely used application of diffraction theory: electromagnetic waves in a vacuum or in a medium with a homogeneous dielectric constant. In this case ψ can be regarded as one of the Cartesian components of the electric field vector, such as E_x (or equally well, a Cartesian component of the vector potential or the magnetic field vector). In a vacuum or in a homogeneous dielectric medium, Maxwell's equations imply that this $\psi = E_x$ satisfies the scalar wave equation exactly and hence, for fixed frequency, the Helmholtz equation (8.1b). However, when the wave hits a boundary of the medium (e.g., the edge of an aperture, or the surface of a mirror or lens), its interaction with the boundary can couple the various components of $\mathbf{E}$, thereby invalidating the simple scalar theory we develop in this chapter. Fortunately, this polarizational coupling is usually weak in the paraxial (small angle) limit, and also under a variety of other circumstances, thereby making our simple scalar formalism quite accurate.[1]

applications of scalar diffraction formalism

---

1. For a formulation of diffraction that takes account of these polarization effects, see, e.g., Born and Wolf (1999, Chap. 11).

The Helmholtz equation (8.1b) is an elliptic, linear, partial differential equation, and we can thus express the value $\psi_{\mathcal{P}}$ of ψ at any point $\mathcal{P}$ inside some closed surface $\mathcal{S}$ as an integral over $\mathcal{S}$ of some linear combination of ψ and its normal derivative; see Fig. 8.1. To derive such an expression, we proceed as follows. First, we introduce in the interior of $\mathcal{S}$ a second solution of the Helmholtz equation:

spherical wave

$$\psi_0 = \frac{e^{ikr}}{r}. \tag{8.2}$$

This is a spherical wave originating from the point $\mathcal{P}$, and r is the distance from $\mathcal{P}$ to the point where ψ_0 is evaluated. Next we apply Gauss's theorem, Eq. (1.28a), to the vector field $\psi \nabla \psi_0 - \psi_0 \nabla \psi$ and invoke Eq. (8.1b), thereby obtaining

$$\int_{\mathcal{S} \cup \mathcal{S}_o} (\psi \nabla \psi_0 - \psi_0 \nabla \psi) \cdot d\mathbf{\Sigma} = -\int_{\mathcal{V}} (\psi \nabla^2 \psi_0 - \psi_0 \nabla^2 \psi) dV$$

$$= 0. \tag{8.3}$$

Here we have introduced a small sphere $\mathcal{S}_o$ of radius r_o surrounding $\mathcal{P}$ (Fig. 8.1); $\mathcal{V}$ is the volume between the two surfaces $\mathcal{S}_o$ and $\mathcal{S}$ (so $\mathcal{S} \cup \mathcal{S}_o$ is the boundary $\partial \mathcal{V}$ of $\mathcal{V}$). For future convenience we have made an unconventional choice of direction for the integration element $d\mathbf{\Sigma}$: it points into $\mathcal{V}$ instead of outward, thereby producing the minus sign in the second expression in Eq. (8.3). As we let the radius r_o decrease to zero, we find that $\psi \nabla \psi_0 - \psi_0 \nabla \psi \to -\psi_{\mathcal{P}}/r_o^2 \, \mathbf{e}_r + O(1/r_o)$, and so the integral over $\mathcal{S}_o$ becomes $-4\pi \psi_{\mathcal{P}}$ (where $\psi_{\mathcal{P}}$ is the value of ψ at $\mathcal{P}$). Rearranging, we obtain

Helmholtz-Kirchhoff integral

$$\boxed{\psi_{\mathcal{P}} = \frac{1}{4\pi} \int_{\mathcal{S}} \left(\psi \nabla \frac{e^{ikr}}{r} - \frac{e^{ikr}}{r} \nabla \psi \right) \cdot d\mathbf{\Sigma}.} \tag{8.4}$$

Equation (8.4), known as the *Helmholtz-Kirchhoff integral*, is the promised expression for the field ψ at some point $\mathcal{P}$ in terms of a linear combination of its value and normal derivative on a surrounding surface. The specific combination of ψ and $d\mathbf{\Sigma} \cdot \nabla \psi$ that appears in this formula is perfectly immune to contributions from any wave that might originate at $\mathcal{P}$ and pass outward through $\mathcal{S}$ (any outgoing wave). The integral thus is influenced only by waves that enter $\mathcal{V}$ through $\mathcal{S}$, propagate through $\mathcal{V}$, and then leave through $\mathcal{S}$. (There cannot be sources inside $\mathcal{S}$, except conceivably at $\mathcal{P}$, because we assumed ψ satisfies the source-free Helmholtz equation throughout $\mathcal{V}$.) If $\mathcal{P}$ is many wavelengths away from the boundary $\mathcal{S}$, then to high accuracy, the integral is influenced by the waves ψ only when they are entering through $\mathcal{S}$ (when they are incoming) and not when they are leaving (outgoing). This fact is important for applications, as we shall see.

8.2.1 Diffraction by an Aperture

Now let us suppose that some aperture $\mathcal{Q}$, with size much larger than a wavelength but much smaller than the distance to $\mathcal{P}$, is illuminated by a distant wave source (left side

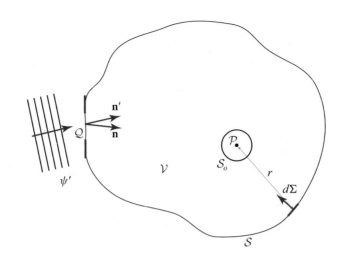

FIGURE 8.1 Geometry for the Helmholtz-Kirchhoff integral (8.4), which expresses the field ψ_P at the point P in terms of an integral of the field and its normal derivative over the surrounding surface S. The small sphere S_o is centered on the observation point P, and V is the volume bounded by S and S_o. (The aperture Q, the vectors $\mathbf{n}$ and $\mathbf{n}'$ at the aperture, the incoming wave ψ', and the point P' are irrelevant to the formulation of the Helmholtz-Kirchhoff integral, but appear in applications later in this chapter—initially in Sec. 8.2.1.)

of Fig. 8.1).[2] Let the surface S pass through the aperture Q, and denote by ψ' the wave incident on Q. We assume that the diffracting aperture has a local and linear effect on ψ'. More specifically, we suppose that the wave transmitted through the aperture is given by

$$\psi_Q = \mathsf{t}\,\psi',\tag{8.5}$$

where t is a complex transmission function that varies over the aperture. In practice, t is usually zero (completely opaque region) or unity (completely transparent region). However, t can also represent a variable phase factor when, for example, the aperture consists of a medium (lens) of variable thickness and of different refractive index from that of the homogeneous medium outside the aperture—as is the case in microscopes, telescopes, eyes, and other optical devices.

complex transmission function t

What this formalism does not allow is that ψ_Q at any point on the aperture be influenced by the wave's interaction with other parts of the aperture. For this reason, not only the aperture, but also any structure that it contains must be many wavelengths across. To give a specific example of what might go wrong, suppose that electromagnetic radiation is normally incident on a wire grid. A surface current will

limitations of this scalar diffraction formalism

---

2. If the aperture were comparable to a wavelength in size, or if part of it were only a few wavelengths from P, then polarizational coupling effects at the aperture would be large (Born and Wolf, 1999). Our assumption avoids this complication.

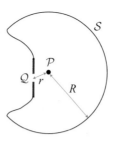

FIGURE 8.2 Geometry for deriving Eq. (8.6) for diffraction by an aperture.

be induced in each wire by the wave's electric field, and that current will produce a secondary wave that cancels the primary wave immediately behind the wire, thereby "eclipsing" the wave. If the wire separation is not large compared to a wavelength, then the secondary wave from the current flowing in one wire will drive currents in adjacent wires, thereby modifying their secondary waves and disturbing the eclipse in a complex, polarization-dependent manner. Such modifications are negligible if the wires are many wavelengths apart.

Let us now use the Helmholtz-Kirchhoff integral (8.4) to compute the field at $\mathcal{P}$ due to the wave $\psi_Q = \mathfrak{t}\,\psi'$ transmitted through the aperture. Let the (imaginary) surface $\mathcal{S}$ in Fig. 8.1 comprise the aperture $\mathcal{Q}$, a sphere of radius $R \gg r$ centered on $\mathcal{P}$, and an extension of the aperture to meet the sphere as sketched in Fig. 8.2. Assume that the only incoming waves are those passing through the aperture. Then, for the reason discussed in the paragraph following Eq. (8.4), when the incoming waves subsequently pass on outward through the spherical part of $\mathcal{S}$, they contribute negligibly to the integral (8.4). Also, with the extension from aperture to sphere swinging back as drawn (Fig. 8.2), the contribution from the extension will also be negligible. Therefore, the only contribution is from the aperture itself.[3]

Because $kr \gg 1$, we can write $\nabla(e^{ikr}/r) \simeq -ik\mathbf{n}e^{ikr}/r$ on the aperture. Here $\mathbf{n}$ is a unit vector pointing toward $\mathcal{P}$ (Fig. 8.1). Similarly, we write $\nabla\psi \simeq ik\mathfrak{t}\,\mathbf{n}'\psi'$, where $\mathbf{n}'$ is a unit vector along the direction of propagation of the incident wave (and where our assumption that anything in the aperture varies on scales long compared to $\lambda = 1/k$ permits us to ignore the gradient of $\mathfrak{t}$). Inserting these gradients into the Helmholtz-Kirchhoff integral, we obtain

---

3. Actually, the incoming waves will diffract around the edge of the aperture onto the back side of the screen that bounds the aperture (i.e., the side facing $\mathcal{P}$), and this diffracted wave will contribute to the Helmholtz-Kirchhoff integral in a polarization-dependent way (see Born and Wolf, 1999, Chap. 11). However, because the diffracted wave decays along the screen with an e-folding length of order a wavelength, its contribution will be negligible if the aperture is many wavelengths across and $\mathcal{P}$ is many wavelengths away from the edge of the aperture, as we have assumed.

$$\psi_P = -\frac{ik}{2\pi} \int_Q d\Sigma \cdot \left(\frac{\mathbf{n} + \mathbf{n}'}{2}\right) \frac{e^{ikr}}{r} t \, \psi'. \qquad (8.6)$$

wave field behind an
aperture

Equation (8.6) can be used to compute the wave from a small aperture Q at any point P in the far field. It has the form of an integral transform of the incident field variable ψ', where the integral is over the area of the aperture. The kernel of the transform is the product of several factors. The factor $1/r$ guarantees that the wave's energy flux falls off as the inverse square of the distance to the aperture, as we might have expected. The phase factor $-ie^{ikr}$ advances the phase of the wave by an amount equal to the optical path length between the element $d\Sigma$ of the aperture and P, minus $\pi/2$ (the phase of $-i$). The amplitude and phase of the wave ψ' can also be changed by the transmission function t. Finally there is the geometric factor $d\hat{\Sigma} \cdot (\mathbf{n} + \mathbf{n}')/2$ (with $d\hat{\Sigma}$ the unit vector normal to the aperture). This is known as the *obliquity factor*, and it ensures that the waves from the aperture propagate only forward with respect to the original wave and not backward (i.e., not in the direction $\mathbf{n} = -\mathbf{n}'$). More specifically, this factor prevents the backward-propagating secondary wavelets in a Huygens construction from reinforcing one another to produce a back-scattered wave. When dealing with paraxial Fourier optics (Sec. 8.5), we can usually set the obliquity factor to unity.

obliquity factor

It is instructive to specialize to a point source seen through a small diffracting aperture. If we suppose that the source has unit strength and is located at P', a distance r' before Q (Fig. 8.1), then $\psi' = -e^{ikr'}/(4\pi r')$ (our definition of unit strength), and ψ_P can be written in the symmetric form

$$\psi_P = \int_Q \left(\frac{e^{ikr}}{4\pi r}\right) i t \, (\mathbf{k} + \mathbf{k}') \cdot d\Sigma \left(\frac{e^{ikr'}}{4\pi r'}\right). \qquad (8.7)$$

propagator through an
aperture

We can think of this expression as the Green's function response at P to a delta function source at P'. Alternatively, we can regard it as a *propagator* from P' to P by way of the aperture (co-opting a concept that was first utilized in quantum mechanics but is also useful in classical optics).

8.2.2 Spreading of the Wavefront: Fresnel and Fraunhofer Regions

8.2.2

Equation (8.6) [or Eq. (8.7)] gives a general prescription for computing the diffraction pattern from an illuminated aperture. It is commonly used in two complementary limits, called "Fraunhofer" and "Fresnel."

Suppose that the aperture has linear size a (as in Fig. 8.3) and is roughly centered on the geometric ray from the source point P' to the field point P, so $\mathbf{n} \cdot d\hat{\Sigma} = \mathbf{n}' \cdot d\hat{\Sigma} \simeq 1$. Consider the variations of the phase φ of the contributions to ψ_P that come from various places in the aperture. Using elementary trigonometry, we can estimate that locations on the aperture's opposite edges produce phases at P that differ by $\Delta\varphi =$

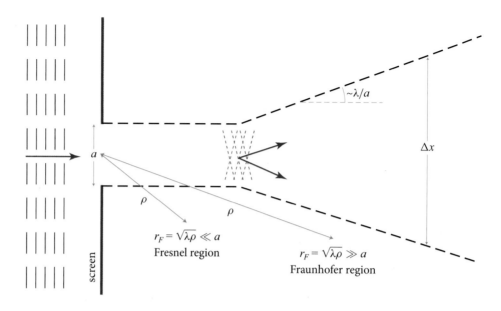

FIGURE 8.3 Fraunhofer and Fresnel diffraction. The dashed line is an approximation to the edge of the aperture's shadow.

$k(\rho_2 - \rho_1) \sim ka^2/2\rho$, where ρ_1 and ρ_2 are the distances of $\mathcal{P}$ from the two edges of the aperture, and ρ is the distance from the center of the aperture. There are two limiting regions for ρ, depending on whether $\mathcal{P}$'s so-called *Fresnel length*,

Fresnel length

$$r_F \equiv \left(\frac{2\pi\rho}{k}\right)^{1/2} = (\lambda\rho)^{1/2} \tag{8.8}$$

regions of Fraunhofer and Fresnel diffraction

(a surrogate for the distance ρ), is large or small compared to the aperture. Notice that $(a/r_F)^2 = ka^2/(2\pi\rho) \sim \Delta\varphi/\pi$. Therefore, when $r_F \gg a$ (field point far from the aperture), the phase variation $\Delta\varphi$ across the aperture is $\ll \pi$ and can be ignored, so the contributions at $\mathcal{P}$ from different parts of the aperture are essentially in phase with one another. This is the *Fraunhofer* region. When $r_F \ll a$ (near the aperture), the phase variation is $\Delta\varphi \gg \pi$ and therefore is of utmost importance in determining the observed energy-flux pattern $F \propto |\psi_\mathcal{P}|^2$. This is the *Fresnel* region; see Fig. 8.3.

We can use an argument familiar, perhaps, from quantum mechanics to deduce the qualitative form of the flux patterns in these two regions. For simplicity, let the incoming wave be planar [r' huge in Eq. (8.7)]; let it propagate perpendicular to the aperture (as shown in Fig. 8.3); and let the aperture be empty, so $t = 1$ inside the aperture. Then geometric optics (photons treated like classical particles) would predict that the aperture's edge will cast a sharp shadow; the wave leaves the plane of the aperture as a beam with a sharp edge. However, wave optics insists that the transverse localization of the wave into a region of size $\Delta x \sim a$ must produce a spread in its transverse wave vector, $\Delta k_x \sim 1/a$ (a momentum uncertainty $\Delta p_x = \hbar\Delta k_x \sim \hbar/a$ in the language of the Heisenberg uncertainty principle). This uncertain transverse wave vector produces, after propagating a distance ρ, a corresponding uncertainty $(\Delta k_x/k)\rho \sim r_F^2/a$ in the beam's transverse size. This uncertainty superposes incoherently on the aperture-induced size a of the beam to produce a net transverse beam size of

$$\Delta x \sim \sqrt{a^2 + (r_F^2/a)^2}$$

$$\sim \begin{cases} a & \text{if } r_F \ll a \text{ (i.e., } \rho \ll a^2/\lambda; \text{ Fresnel region),} \\ \left(\dfrac{\lambda}{a}\right)\rho & \text{if } r_F \gg a \text{ (i.e., } \rho \gg a^2/\lambda; \text{ Fraunhofer region).} \end{cases} \tag{8.9}$$

Therefore, in the nearby, Fresnel region, the aperture creates a beam whose edges have the same shape and size as the aperture itself, and are reasonably sharp (but with some oscillatory blurring, associated with wavepacket spreading, that we analyze below); see Fig. 8.4. Thus, in the Fresnel region the field behaves approximately as one would predict using geometric optics. By contrast, in the more distant, Fraunhofer region, wavefront spreading causes the transverse size of the entire beam to grow linearly with distance; as illustrated in Fig. 8.4, the flux pattern differs markedly from the aperture's shape. We analyze the distant (Fraunhofer) region in Sec. 8.3 and the near (Fresnel) region in Sec. 8.4.

Fresnel and Fraunhofer diffraction regions and patterns

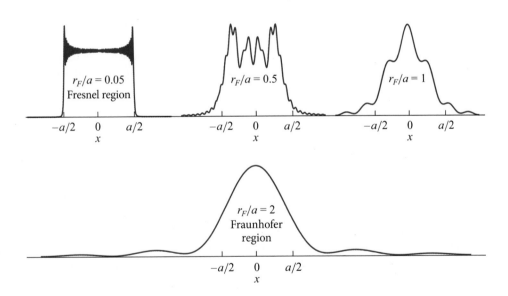

FIGURE 8.4 The 1-dimensional energy-flux diffraction pattern $|\psi|^2$ produced on a screen a distance z from a slit with $t(x) = 1$ for $|x| < a/2$ and $t(x) = 0$ for $|x| > a/2$. Four patterns are shown, each for a different value of $r_F/a = \sqrt{\lambda z}/a$. For $r_F/a = 0.05$ (very near the slit; extreme Fresnel region), the flux distribution resembles the slit itself: sharp edges at $x = \pm a/2$, but with damped oscillations (interference fringes) near the edges. For $r_F/a = 1$ (beginning of Fraunhofer region) there is a bright central peak and low-brightness, oscillatory side bands. As r_F/a increases from 0.05 to 2, the pattern transitions (quite rapidly between 0.5 and 2) from the Fresnel to the Fraunhofer pattern. These flux distributions are derived in Ex. 8.8.

8.3 Fraunhofer Diffraction

Consider the Fraunhofer region of strong wavefront spreading, $r_F \gg a$, and for simplicity specialize to the case of an incident plane wave with wave vector $\mathbf{k}$ orthogonal to the aperture plane; see Fig. 8.5. Regard the line along $\mathbf{k}$ through the center of the aperture $\mathcal{Q}$ as the optic axis and identify points in the aperture by their transverse 2-dimensional vectorial separation $\mathbf{x}$ from that axis. Identify $\mathcal{P}$ by its distance ρ from the aperture center and its 2-dimensional transverse separation $\rho\boldsymbol{\theta}$ from the optic axis, and restrict attention to small-angle diffraction $|\boldsymbol{\theta}| \ll 1$. Then the geometric path length between $\mathcal{P}$ and a point $\mathbf{x}$ on $\mathcal{Q}$ [the length denoted r in Eq. (8.6)] can be expanded as

$$\text{Path length} = r = (\rho^2 - 2\rho\mathbf{x} \cdot \boldsymbol{\theta} + x^2)^{1/2} \simeq \rho - \mathbf{x} \cdot \boldsymbol{\theta} + \frac{x^2}{2\rho} + \ldots; \quad (8.10)$$

cf. Fig. 8.5. The first term in this expression, ρ, just contributes an $\mathbf{x}$-independent phase $e^{ik\rho}$ to the $\psi_{\mathcal{P}}$ of Eq. (8.6). The third term, $x^2/(2\rho)$, contributes a phase variation that is $\ll 1$ in the Fraunhofer region (but that will be important in the Fresnel region; Sec. 8.4). Therefore, in the Fraunhofer region we can retain just the second term, $-\mathbf{x} \cdot \boldsymbol{\theta}$, and write Eq. (8.6) in the form

Fraunhofer diffraction of a plane wave

$$\psi_{\mathcal{P}}(\boldsymbol{\theta}) \propto \int e^{-ik\mathbf{x}\cdot\boldsymbol{\theta}} \mathfrak{t}(\mathbf{x})\, d\Sigma \equiv \tilde{\mathfrak{t}}(\boldsymbol{\theta}), \quad (8.11a)$$

where $d\Sigma$ is the surface-area element in the aperture plane, and we have dropped a constant phase factor and constant multiplicative factors. Thus, $\psi_{\mathcal{P}}(\boldsymbol{\theta})$ in the Fraunhofer region is given by the 2-dimensional Fourier transform, denoted $\tilde{\mathfrak{t}}(\boldsymbol{\theta})$, of the transmission function $\mathfrak{t}(\mathbf{x})$, with $\mathbf{x}$ made dimensionless in the transform by multiplying by $k = 2\pi/\lambda$. [Note that with this optics convention for the Fourier transform, the inverse transform is

optics convention for Fourier transform

$$\mathfrak{t}(\mathbf{x}) = \left(\frac{k}{2\pi}\right)^2 \int e^{ik\mathbf{x}\cdot\boldsymbol{\theta}} \tilde{\mathfrak{t}}(\boldsymbol{\theta})\, d\Omega, \quad (8.11b)$$

where $d\Omega = d\theta_x d\theta_y$ is the solid angle.]

The flux distribution of the diffracted wave [Eq. (8.11a)] is

Fraunhofer diffracted flux

$$F(\boldsymbol{\theta}) = \overline{\left(\Re[\psi_{\mathcal{P}}(\boldsymbol{\theta})e^{-i\omega t}]\right)^2} = \tfrac{1}{2}|\psi_{\mathcal{P}}(\boldsymbol{\theta})|^2 \propto |\tilde{\mathfrak{t}}(\boldsymbol{\theta})|^2, \quad (8.12)$$

where $\Re$ means take the real part, and the bar means average over time.

As an example, the bottom curve in Fig. 8.4 (the curve $r_F = 2a$) shows the flux distribution $F(\theta)$ from a slit

$$\mathfrak{t}(x) = H_1(x) \equiv \begin{cases} 1 & |x| < a/2 \\ 0 & |x| > a/2, \end{cases} \quad (8.13a)$$

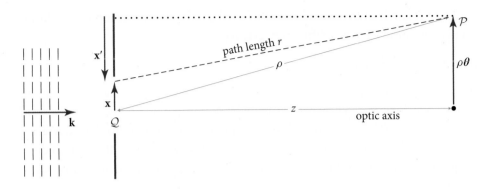

FIGURE 8.5 Geometry for computing the path length between a point $\mathcal{Q}$ in the aperture and the point of observation $\mathcal{P}$. The transverse vector $\mathbf{x}$ is used to identify $\mathcal{Q}$ in our Fraunhofer analysis (Sec. 8.3), and $\mathbf{x}'$ is used in the Fresnel analysis (Sec. 8.4).

for which

$$\psi_{\mathcal{P}}(\theta) \propto \tilde{H}_1 \propto \int_{-a/2}^{a/2} e^{ikx\theta} dx \propto \operatorname{sinc}\left(\tfrac{1}{2}ka\theta\right) , \tag{8.13b}$$

$$F(\theta) \propto \operatorname{sinc}^2\left(\tfrac{1}{2}ka\theta\right) . \tag{8.13c}$$

Fraunhofer diffraction from a slit with width a

Here $\operatorname{sinc}(\xi) \equiv \sin(\xi)/\xi$. The bottom flux curve in Fig. 8.4 is almost—but not quite—described by Eq. (8.13c); the differences (e.g., the not-quite-zero value of the minimum between the central peak and the first side lobe) are due to the field point not being fully in the Fraunhofer region, $r_F/a = 2$ rather than $r_F/a \gg 1$.

It is usually uninteresting to normalize a Fraunhofer diffraction pattern. On those rare occasions when the absolute value of the observed flux is needed, rather than just the pattern's angular shape, it typically can be derived most easily from conservation of the total wave energy. This is why we have ignored the proportionality factors in the diffraction patterns.

All techniques for handling Fourier transforms (which should be familiar from quantum mechanics and elsewhere, though with different normalizations) can be applied to derive Fraunhofer diffraction patterns. In particular, the *convolution theorem* turns out to be very useful. It says that the Fourier transform of the convolution

$$f_2 \otimes f_1 \equiv \int_{-\infty}^{+\infty} f_2(\mathbf{x} - \mathbf{x}') f_1(\mathbf{x}') d\Sigma' \tag{8.14}$$

convolution

of two functions f_1 and f_2 is equal to the product $\tilde{f}_2(\boldsymbol{\theta})\tilde{f}_1(\boldsymbol{\theta})$ of their Fourier transforms, and conversely, the Fourier transform of a product is equal to the convolution. [Here and throughout this chapter, we use the optics version of a Fourier transform, in which 2-dimensional transverse position $\mathbf{x}$ is made dimensionless via the wave number k; Eq. (8.11a).]

As an example of the convolution theorem's power, we now compute the diffraction pattern produced by a diffraction grating.

8.3.1 Diffraction Grating

A diffraction grating[4] can be modeled as a finite series of alternating transparent and opaque, long, parallel strips. Let there be N transparent and opaque strips each of width $a \gg \lambda$ (Fig. 8.6a), and idealize them as infinitely long, so their diffraction pattern is 1-dimensional. We outline how to use the convolution theorem to derive the grating's Fraunhofer diffraction pattern. The details are left as an exercise for the reader (Ex. 8.2).

using convolution to build transmission function

Our idealized N-slit grating can be regarded as an infinite series of delta functions with separation $2a$, convolved with the transmission function H_1 [Eq. (8.13a)] for a single slit of width a,

$$\int_{-\infty}^{\infty} \left[\sum_{n=-\infty}^{+\infty} \delta(\xi - 2an) \right] H_1(x - \xi)d\xi, \tag{8.15a}$$

and then multiplied by an aperture function with width $2Na$:

$$H_{2N}(x) \equiv \begin{cases} 1 & |x| < Na \\ 0 & |x| > Na. \end{cases} \tag{8.15b}$$

More explicitly, we have

$$\mathfrak{t}(x) = \left(\int_{-\infty}^{\infty} \left[\sum_{n=-\infty}^{+\infty} \delta(\xi - 2an) \right] H_1(x - \xi)d\xi \right) H_{2N}(x), \tag{8.16}$$

which is shown graphically in Fig. 8.6b.

using the convolution theorem to compute Fraunhofer diffraction

Let us use the convolution theorem to evaluate the Fourier transform $\tilde{\mathfrak{t}}(\theta)$ of expression (8.16), thereby obtaining the diffraction pattern $\psi_{\mathcal{P}}(\theta) \propto \tilde{\mathfrak{t}}(\theta)$ for our transmission grating. The Fourier transform of the infinite series of delta functions with spacing $2a$ is itself an infinite series of delta functions with reciprocal spacing $2\pi/(2ka) = \lambda/(2a)$ (see the hint in Ex. 8.2). This must be multiplied by the Fourier transform $\tilde{H}_1(\theta) \propto \mathrm{sinc}(\frac{1}{2}ka\theta)$ of the single narrow slit and then convolved with the Fourier transform $\tilde{H}_{2N}(\theta) \propto \mathrm{sinc}(Nka\theta)$ of the aperture (wide slit). The result is shown schematically in Fig. 8.6c. (Each of the transforms is real, so the 1-dimensional functions shown in the figure fully embody them.)

The resulting diffracted energy flux, $F \propto |\mathfrak{t}(\theta)|^2$ (as computed in Ex. 8.2), is shown in Fig. 8.6d. The grating has channeled the incident radiation into a few equally spaced beams with directions $\theta = \pi p/(ka) = p\lambda/(2a)$, where p is an integer known as the *order* of the beam. Each of these beams has a shape given by $|\tilde{H}_{2N}(\theta)|^2$: a sharp central peak with half-width (distance from center of peak to first null of the flux) $\lambda/(2Na)$, followed by a set of *side lobes* whose intensities are $\propto N^{-1}$.

---

4. Diffraction gratings were first fabricated by David Rittenhouse in 1785 by winding hair around a pair of screws.

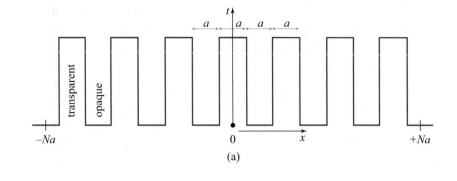

(a)

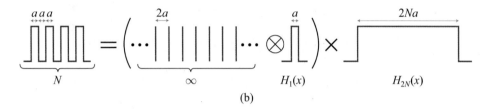

(b)

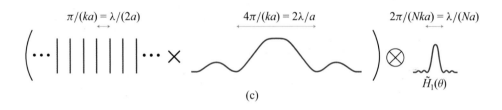

(c)

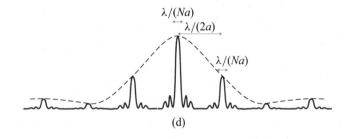

(d)

FIGURE 8.6 (a) Diffraction grating $t(x)$ formed by N alternating transparent and opaque strips each of width a. (b) Decomposition of this finite grating into an infinite series of equally spaced delta functions that are convolved (the symbol $\otimes$) with the shape of an individual transparent strip (i.e., a slit) and then multiplied (the symbol $\times$) by a large aperture function covering N such slits; Eq. (8.16). (c) The resulting Fraunhofer diffraction pattern $\tilde{t}(\theta)$ is shown schematically as the Fourier transform of a series of delta functions multiplied by the Fourier transform of a single slit and then convolved with the Fourier transform of the aperture. (d) The energy flux $F \propto |\tilde{t}(\theta)|^2$ of this diffraction pattern.

The deflection angles $\theta = p\lambda/(2a)$ of these beams are proportional to λ; this underpins the use of diffraction gratings for spectroscopy (different wavelengths deflected into beams at different angles). It is of interest to ask what the wavelength resolution of such an idealized grating might be. If one focuses attention on the pth-order beams at two wavelengths λ and $\lambda + \delta\lambda$ [which are located at $\theta = p\lambda/(2a)$ and $p(\lambda + \delta\lambda)/(2a)$], then one can distinguish the beams from each other when their separation $\delta\theta = p\delta\lambda/(2a)$ is at least as large as the angular distance $\lambda/(2Na)$ between the maximum of each beam's diffraction pattern and its first minimum, that is, when

$$\frac{\lambda}{\delta\lambda} \lesssim \mathcal{R} \equiv Np. \tag{8.17}$$

(Recall that N is the total number of slits in the grating, and p is the order of the diffracted beam.) This $\mathcal{R}$ is called the grating's *chromatic resolving power*.

Real gratings are not this simple. First, they usually work not by modulating the amplitude of the incident radiation in this simple manner, but instead by modulating the phase. Second, the manner in which the phase is modulated is such as to channel most of the incident power into a particular order, a technique known as *blazing*. Third, gratings are often used in reflection rather than transmission. Despite these complications, the principles of a real grating's operation are essentially the same as our idealized grating. Manufactured gratings typically have $N \gtrsim 10{,}000$, giving a wavelength resolution for visual light that can be as small as $\lambda/10^5 \sim 10$ pm (i.e., 10^{-11} m).

EXERCISES

Exercise 8.1 *Practice: Convolutions and Fourier Transforms*

(a) Calculate the 1-dimensional Fourier transforms [Eq. (8.11a) reduced to 1 dimension] of the functions $f_1(x) \equiv e^{-x^2/2\sigma^2}$, and $f_2 \equiv 0$ for $x < 0$, $f_2 \equiv e^{-x/h}$ for $x \geq 0$.

(b) Take the inverse transforms of your answers to part (a) and recover the original functions.

(c) Convolve the exponential function f_2 with the Gaussian function f_1, and then compute the Fourier transform of their convolution. Verify that the result is the same as the product of the Fourier transforms of f_1 and f_2.

Exercise 8.2 *Derivation: Diffraction Grating*

Use the convolution theorem to carry out the calculation of the Fraunhofer diffraction pattern from the grating shown in Fig. 8.6. [Hint: To show that the Fourier transform of the infinite sequence of equally spaced delta functions is a similar sequence of delta functions, perform the Fourier transform to get $\sum_{n=-\infty}^{+\infty} e^{i2kan\theta}$ (aside from a multiplicative factor); then use the formulas for a Fourier *series* expansion, and its inverse, for any function that is periodic with period $\pi/(ka)$ to show that $\sum_{n=-\infty}^{+\infty} e^{i2kan\theta}$ is a sequence of delta functions.]

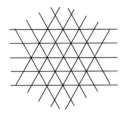

FIGURE 8.7 Diffraction grating formed
from three groups of parallel lines.

Exercise 8.3 *Problem: Triangular Diffraction Grating*
Sketch the Fraunhofer diffraction pattern you would expect to see from a diffraction grating made from three groups of parallel lines aligned at angles of 120° to one another (Fig. 8.7).

8.3.2 Airy Pattern of a Circular Aperture: Hubble Space Telescope

The Hubble Space Telescope was launched in April 1990 to observe planets, stars, and galaxies above Earth's atmosphere. One reason for going into space is to avoid the irregular refractive-index variations in Earth's atmosphere, known, generically, as *seeing,* which degrade the quality of the images. (Another reason is to observe the ultraviolet part of the spectrum, which is absorbed in Earth's atmosphere.) Seeing typically limits the angular resolution of Earth-bound telescopes to ~0.5″ at visual wavelengths (see Box 9.2). We wish to compute how much the angular resolution improves by going into space. As we shall see, the computation is essentially an exercise in Fraunhofer diffraction theory.

The essence of the computation is to idealize the telescope as a circular aperture with diameter equal to that of the primary mirror. Light from this mirror is actually reflected onto a secondary mirror and then follows a complex optical path before being focused on a variety of detectors. However, this path is irrelevant to the angular resolution. The purposes of the optics are merely (i) to bring the Fraunhofer-region light to a focus close to the mirror [Eq. (8.31) and subsequent discussion], in order to produce an instrument that is compact enough to be launched, and (ii) to match the sizes of stars' images to the pixel size on the detector. In doing so, however, the optics leaves the angular resolution unchanged; the resolution is the same as if we were to observe the light, which passes through the primary mirror's circular aperture, far beyond the mirror, in the Fraunhofer region.

If the telescope aperture were very small, for example, a pin hole, then the light from a point source (a very distant star) would create a broad diffraction pattern, and the telescope's angular resolution would be correspondingly poor. As we increase the diameter of the aperture, we still see a diffraction pattern, but its angular width diminishes.

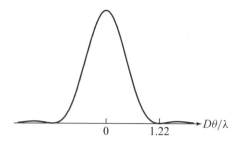

FIGURE 8.8 Airy diffraction pattern produced by a circular aperture.

Using these considerations, we can compute how well the telescope can distinguish neighboring stars. We do not expect it to resolve them fully if they are closer together on the sky than the angular width of the diffraction patttern. Of course, optical imperfections and pointing errors in a real telescope may degrade the image quality even further, but this is the best that can be done, limited only by the uncertainty principle.

The calculation of the Fraunhofer amplitude far from the aperture, via the Fourier transform (8.11a), is straightforward (Ex. 8.4):

Airy diffraction pattern from circular aperture

$$\psi(\theta) \propto \int_{\text{Disk with diameter } D} e^{-ik\mathbf{x}\cdot\boldsymbol{\theta}} d\Sigma$$

$$\propto \text{jinc}\left(\frac{kD\theta}{2}\right), \tag{8.18}$$

where D is the diameter of the aperture (i.e., of the telescope's primary mirror), $\theta \equiv |\boldsymbol{\theta}|$ is the angle from the optic axis, and $\text{jinc}(x) \equiv J_1(x)/x$, with J_1 the Bessel function of the first kind and of order one. The flux from the star observed at angle θ is therefore $\propto \text{jinc}^2(kD\theta/2)$. This energy-flux pattern, known as the *Airy pattern,* is shown in Fig. 8.8. The image appears as a central "Airy disk" surrounded by a circle where the flux vanishes, and then further surrounded by a series of concentric rings whose flux diminishes with radius. Only 16% of the total light falls outside the central Airy disk. The angular radius θ_A of the Airy disk (i.e., the radius of the dark circle surrounding it) is determined by the first zero of $J_1(kD\theta/2) = J_1(\theta \pi D/\lambda)$:

$$\boxed{\theta_A = 1.22\lambda/D.} \tag{8.19}$$

A conventional, though essentially arbitrary, criterion for angular resolution is to say that two point sources can be distinguished if they are separated in angle by more than θ_A. For the Hubble Space Telescope, $D = 2.4$ m and $\theta_A \sim 0.05''$ at visual wavelengths, which is more than 10 times better than is achievable on the ground with conventional (nonadaptive) optics.

Initially, there was a serious problem with Hubble's telescope optics. The hyperboloidal primary mirror was ground to the wrong shape, so rays parallel to the optic axis did not pass through a common focus after reflection off a convex hyperboloidal secondary mirror. This defect, known as *spherical aberration* (Sec. 7.5.2), created blurred images. However, it was possible to correct this error in subsequent instruments in the optical train, and the Hubble Space Telescope became the most successful optical telescope of all time, transforming our view of the universe.

spherical aberration in the Hubble Space Telescope

The Hubble Space Telescope should be succeeded in 2018 by the James Webb Space Telescope. Webb will have a diameter $D \simeq 6.5$ m (2.7 times larger than Hubble) but its wavelengths of observation are somewhat longer (0.6 μm to 28.5 μm), so its angular resolution will be, on average, roughly the same as Hubble's.

EXERCISES

Exercise 8.4 *Derivation: Airy Pattern*
Derive and plot the Airy diffraction pattern [Eq. (8.18)] and show that 84% of the light is contained within the Airy disk.

Exercise 8.5 *Problem: Pointillist Painting*
The neoimpressionist painter Georges Seurat was a member of the pointillist school. His paintings consisted of an enormous number of closely spaced dots of pure pigment (of size ranging from ~0.4 mm in his smaller paintings to ~4 mm in his largest paintings, such as *Gray Weather, Grande Jatte,* Fig. 8.9). The illusion of color mixing was produced only in the eye of the observer. How far from the painting should one stand to obtain the desired blending of color?

(a)   (b)

FIGURE 8.9 (a) Georges Seurat's painting *Gray Weather, Grande Jatte*. When viewed from a sufficient distance, adjacent dots of paint with different colors blend together in the eye to form another color. (b) Enlargement of the boat near the center of the painting. In this enlargement, one sees clearly the individual dots of paint. *Gray Weather, Grande Jatte* by Georges Seurat (ca. 1886–88); The Metropolitan Museum of Art (www.metmuseum.org); The Walter H. and Leonore Annenberg Collection, Gift of Walter H. and Leonore Annenberg, 2002, Bequest of Walter H. Annenberg, 2002.

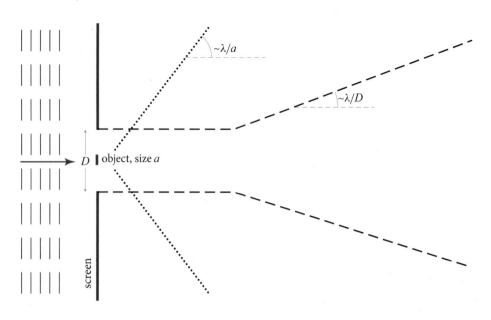

FIGURE 8.10 Geometry for Babinet's principle. The beam produced by the large aperture D is confined between the long-dashed lines. Outside this beam, the energy-flux pattern $F(\boldsymbol{\theta}) \propto |\mathfrak{t}(\boldsymbol{\theta})|^2$ produced by a small object (size a) and its complement are the same, Eqs. (8.20). This diffracted flux pattern is confined between the dotted lines.

8.3.3 8.3.3 Babinet's Principle

Suppose that monochromatic light falls normally onto a large circular aperture with diameter D. At distances $z \lesssim D^2/\lambda$ (i.e., $r_F \lesssim D$), the transmitted light will be collimated into a beam with diameter D, and at larger distances, the beam will become conical with opening angle λ/D (Fig. 8.3) and flux distribution given by the Airy diffraction pattern of Fig. 8.8.

Now place into this aperture a significantly smaller object (size $a \ll D$; Fig. 8.10) with transmissivity $\mathfrak{t}_1(\mathbf{x})$—for example, an opaque star-shaped object. This object will produce a Fraunhofer diffraction pattern with opening angle $\lambda/a \gg \lambda/D$ that extends well beyond the large aperture's beam. Outside that beam, the diffraction pattern will be insensitive to the shape and size of the large aperture, because only the small object can diffract light to these large angles; so the diffracted flux will be $F_1(\boldsymbol{\theta}) \propto |\tilde{\mathfrak{t}}_1(\theta)|^2$.

Next, suppose that we replace the small object by one with a *complementary transmissivity* $\mathfrak{t}_2$, complementary in the sense that

$$\boxed{\mathfrak{t}_1(\mathbf{x}) + \mathfrak{t}_2(\mathbf{x}) = 1.} \tag{8.20a}$$

Babinet's principle

For example, we replace a small, opaque star-shaped object by an opaque screen that fills the original, large aperture except for a star-shaped hole. This complementary object will produce a diffraction pattern $F_2(\boldsymbol{\theta}) \propto |\tilde{\mathfrak{t}}_2(\theta)|^2$. Outside the large aperture's beam, this pattern again is insensitive to the size and shape of the large aperture, that is, it is insensitive to the 1 in $\mathfrak{t}_2 = 1 - \mathfrak{t}_1$ (which sends light solely inside the large aperture's beam); so at these large angles, $\tilde{t}_2(\boldsymbol{\theta}) = -\tilde{t}_1(\boldsymbol{\theta})$, which implies that the energy-flux diffraction pattern of the original object and the new, complementary object will be the same outside the large aperture's beam:

$$\boxed{F_2(\boldsymbol{\theta}) \propto |\tilde{t}_2(\boldsymbol{\theta})|^2 = |\tilde{t}_1(\boldsymbol{\theta})|^2 \propto F_1(\boldsymbol{\theta}).}$$

(8.20b)

Exercise 8.6 *Problem: Light Scattering by a Large, Opaque Particle*
Consider the scattering of light by an opaque particle with size $a \gg 1/k$. Neglect any scattering via excitation of electrons in the particle. Then the scattering is solely due to diffraction of light around the particle. With the aid of Babinet's principle, do the following.

(a) Explain why the scattered light is confined to a cone with opening angle $\Delta\theta \sim \pi/(ka) \ll 1$.

(b) Show that the total power in the scattered light, at very large distances from the particle, is $P_S = FA$, where F is the incident energy flux, and A is the cross sectional area of the particle perpendicular to the incident wave.

(c) Explain why the total "extinction" (absorption plus scattering) cross section is equal to $2A$ independent of the shape of the opaque particle.

Exercise 8.7 *Problem: Thickness of a Human Hair*
Conceive and carry out an experiment using light diffraction to measure the thickness of a hair from your head, accurate to within a factor of ~ 2. [Hint: Make sure the source of light that you use is small enough—e.g., a very narrow laser beam—that its finite size has negligible influence on your result.]

8.4 Fresnel Diffraction

We next turn to the Fresnel region of observation points $\mathcal{P}$ with $r_F = \sqrt{\lambda\rho}$ much smaller than the aperture. In this region, the field at $\mathcal{P}$ arriving from different parts of the aperture has significantly different phase $\Delta\varphi \gg 1$. We again specialize to incoming wave vectors that are approximately orthogonal to the aperture plane and to small diffraction angles, so we can ignore the obliquity factor $d\boldsymbol{\Sigma} \cdot (\mathbf{n} + \mathbf{n}')/2$ in Eq. (8.6). By contrast with the Fraunhofer case, we now identify $\mathcal{P}$ by its distance z from the aperture plane instead of its distance ρ from the aperture center, and we use as our integration variable in the aperture $\mathbf{x}' \equiv \mathbf{x} - \rho\boldsymbol{\theta}$ (Fig. 8.5), thereby writing the dependence of the phase at $\mathcal{P}$ on $\mathbf{x}$ in the form

$$\Delta\varphi \equiv k \times [(\text{path length from } \mathbf{x} \text{ to } \mathcal{P}) - z] = \frac{k\mathbf{x}'^2}{2z} + O\left(\frac{k x'^4}{z^3}\right).$$

(8.21)

In the Fraunhofer region (Sec. 8.3) only the linear term $-k\mathbf{x} \cdot \boldsymbol{\theta}$ in $k\mathbf{x}'^2/(2z) \simeq k(\mathbf{x} - r\boldsymbol{\theta})^2/r$ was significant. In the Fresnel region the term quadratic in $\mathbf{x}$ is also significant (and we have changed variables to $\mathbf{x}'$ so as to simplify it), but the $O(x'^4)$ term is negligible. Therefore, in the Fresnel region the diffraction pattern (8.6) is

$$\psi_P = \frac{-ike^{ikz}}{2\pi z}\int e^{i\Delta\varphi}\mathfrak{t}\psi'd\Sigma' = \frac{-ike^{ikz}}{2\pi z}\int e^{ik\mathbf{x}'^2/(2z)}\mathfrak{t}(\mathbf{x}')\psi'(\mathbf{x}')d\Sigma', \quad (8.22)$$

where in the denominator we have replaced r by z to excellent approximation.

Let us consider the Fresnel diffraction pattern formed by an empty ($\mathfrak{t} = 1$) simple aperture of arbitrary shape, illuminated by a normally incident plane wave. It is convenient to introduce transverse Cartesian coordinates $\mathbf{x}' = (x', y')$ and define

$$\sigma = \left(\frac{k}{\pi z}\right)^{1/2}x', \qquad \tau = \left(\frac{k}{\pi z}\right)^{1/2}y'. \qquad (8.23a)$$

[Notice that $(k/[\pi z])^{1/2}$ is $\sqrt{2}/r_F$; cf. Eq. (8.8).] We can thereby rewrite Eq. (8.22) in the form

Fresnel diffraction pattern from arbitrary empty aperture $\mathcal{Q}$

$$\psi_P = -\frac{i}{2}\int\int_{\mathcal{Q}}e^{i\pi\sigma^2/2}e^{i\pi\tau^2/2}\psi_{\mathcal{Q}}e^{ikz}d\sigma\,d\tau. \qquad (8.23b)$$

Here we have changed notation for the field ψ' impinging on the aperture $\mathcal{Q}$ to $\psi_{\mathcal{Q}}$.

Equations (8.23) depend on the transverse location of the observation point $\mathcal{P}$ through the origin used to define $\mathbf{x}' = (x', y')$; see Fig. 8.5. We use these rather general expressions in Sec. 8.5, when discussing paraxial Fourier optics, as well as in Secs. 8.4.1–8.4.4 on Fresnel diffraction.

8.4.1 Rectangular Aperture, Fresnel Integrals, and the Cornu Spiral

In this and the following three sections, we explore the details of the Fresnel diffraction pattern for an incoming plane wave that falls perpendicularly on the aperture, so $\psi_{\mathcal{Q}}$ is constant over the aperture.

Fresnel diffraction by rectangular aperture

For simplicity, we initially confine attention to a rectangular aperture with edges along the x' and y' directions. Then the two integrals have limits that are independent of each other, and the integrals can be expressed in the form $\mathcal{E}(\sigma_{max}) - \mathcal{E}(\sigma_{min})$ and $\mathcal{E}(\tau_{max}) - \mathcal{E}(\tau_{min})$, so

$$\psi_P = \frac{-i}{2}[\mathcal{E}(\sigma_{max}) - \mathcal{E}(\sigma_{min})][\mathcal{E}(\tau_{max}) - \mathcal{E}(\tau_{min})]\psi_{\mathcal{Q}}e^{ikz} \equiv \frac{-i}{2}\Delta\mathcal{E}_\sigma\Delta\mathcal{E}_\tau\psi_{\mathcal{Q}}e^{ikz},$$

$$(8.24a)$$

where the arguments are the limits of integration (the two edges of the aperture), and where

$$\mathcal{E}(\xi) \equiv \int_0^\xi e^{i\pi\sigma^2/2}d\sigma \equiv C(\xi) + iS(\xi). \qquad (8.24b)$$

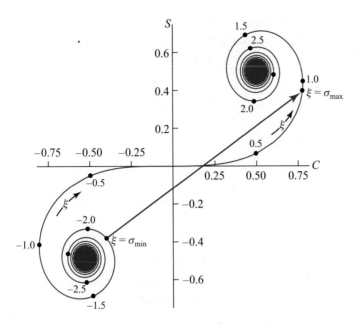

FIGURE 8.11 Cornu spiral in the complex plane; the real part of $\mathcal{E}(\xi) = C(\xi) + iS(\xi)$ is plotted horizontally, and the imaginary part vertically; the point $\xi = 0$ is at the origin, positive ξ is in the upper right quadrant, and negative ξ is in the lower left quadrant. The diffracted energy flux is proportional to the squared length of the arrow reaching from $\xi = \sigma_{min}$ to $\xi = \sigma_{max}$.

Here

$$C(\xi) \equiv \int_0^\xi d\sigma \, \cos(\pi\sigma^2/2), \qquad S(\xi) \equiv \int_0^\xi d\sigma \, \sin(\pi\sigma^2/2) \qquad (8.24c) \qquad \textbf{Fresnel integrals}$$

are known as *Fresnel integrals* and are standard functions tabulated in many books and can be found in Mathematica, Matlab, and Maple. Notice that the energy-flux distribution is

$$F \propto |\psi_\mathcal{P}|^2 \propto |\Delta\mathcal{E}_\sigma|^2 |\Delta\mathcal{E}_\tau|^2. \qquad (8.24d)$$

It is convenient to exhibit the Fresnel integrals graphically using a *Cornu spiral*   **Cornu spiral** (Fig. 8.11). This is a graph of the parametric equation $[C(\xi), S(\xi)]$, or equivalently, a graph of $\mathcal{E}(\xi) = C(\xi) + iS(\xi)$ in the complex plane. The two terms $\Delta\mathcal{E}_\sigma$ and $\Delta\mathcal{E}_\tau$ in Eq. (8.24a) can be represented in amplitude and phase by arrows in the C-S plane reaching from $\xi = \sigma_{min}$ on the Cornu spiral to $\xi = \sigma_{max}$ (Fig. 8.11), and from $\xi = \tau_{min}$ to $\xi = \tau_{max}$. Correspondingly, the flux F [Eq. (8.24d)] is proportional to the product of the squared lengths of these two vectors.

As the observation point $\mathcal{P}$ moves around in the observation plane (Fig. 8.5), x'_{min} and x'_{max} change, and hence $\sigma_{min} = [k/(\pi z)]^{1/2} x'_{min}$ and $\sigma_{max} = [k/(\pi z)]^{1/2} x'_{max}$ change [i.e., the tail and tip of the arrow in Fig. 8.11 move along the Cornu spiral, thereby changing the diffracted flux, which is $\propto$ (length of arrow)^2].

8.4.2 Unobscured Plane Wave

The simplest illustration of Fresnel integrals is the totally unobscured plane wave. In this case, the limits of both integrations extend from $-\infty$ to $+\infty$, which, as we see in Fig. 8.11, is an arrow of length $2^{1/2}$ and phase $\pi/4$. Therefore, $\psi_{\mathcal{P}}$ is equal to $(2^{1/2}e^{i\pi/4})^2(-i/2)\psi_{\mathcal{Q}}e^{ikz} = \psi_{\mathcal{Q}}e^{ikz}$, as we could have deduced simply by solving the Helmholtz equation (8.1b) for a plane wave.

This unobscured-wavefront calculation elucidates three issues that we have already met. First, it illustrates our interpretation of Fermat's principle in geometric optics. In the limit of short wavelengths, the paths that contribute to the wave field are just those along which the phase is stationary to small variations. Our present calculation shows that, because of the tightening of the Cornu spiral as one moves toward a large argument, the paths that contribute significantly to $\psi_{\mathcal{P}}$ are those that are separated from the geometric-optics path by less than a few Fresnel lengths at $\mathcal{Q}$. (For a laboratory experiment with light and $z \sim 2$ m, the Fresnel length is $\sqrt{\lambda z} \sim 1$ mm.)

for Fresnel diffraction dominant paths are near geometric optics path

A second, and related, point is that, when computing the Fresnel diffraction pattern from a more complicated aperture, we need only perform the integral (8.6) in the immediate vicinity of the geometric-optics ray. We can ignore the contribution from the extension of the aperture $\mathcal{Q}$ to meet the "sphere at infinity" (the surface $\mathcal{S}$ in Fig. 8.2), even when the wave is unobstructed there. The rapid phase variation makes the contribution from that extension and from $\mathcal{S}$ sum to zero.

Third, when integrating over the whole area of the wavefront at $\mathcal{Q}$, we have summed contributions with increasingly large phase differences that add in such a way that the total has a net extra phase of $\pi/2$ relative to the geometric-optics ray. This phase factor cancels exactly the prefactor $-i$ in the Helmholtz-Kirchhoff formula (8.6). (This phase factor is unimportant in the limit of geometric optics.)

8.4.3 Fresnel Diffraction by a Straight Edge: Lunar Occultation of a Radio Source

The next-simplest case of Fresnel diffraction is the pattern formed by a straight edge. As a specific example, consider a cosmologically distant source of radio waves that is occulted by the Moon. If we treat the lunar limb as a straight edge, then the radio source will create a changing diffraction pattern as it passes behind the Moon, and the diffraction pattern can be measured by a radio telescope on Earth. We orient our coordinates so the Moon's edge is along the y' direction (τ direction). Then in Eq. (8.24a) $\Delta\mathcal{E}_\tau \equiv \mathcal{E}(\tau_{\max}) - \mathcal{E}(\tau_{\min}) = \sqrt{2i}$ is constant, and $\Delta\mathcal{E}_\sigma \equiv \mathcal{E}(\sigma_{\max}) - \mathcal{E}(\sigma_{\min})$ is described by the Cornu spiral of Fig. 8.12b.

Long before the occultation, we can approximate $\sigma_{\min} = -\infty$ and $\sigma_{\max} = +\infty$ (Fig. 8.12a), so $\Delta\mathcal{E}_\sigma$ is given by the arrow from $(-1/2, -1/2)$ to $(1/2, 1/2)$ in Fig. 8.12b (i.e., $\Delta\mathcal{E}_\sigma = \sqrt{2i}$). The observed wave amplitude, Eq. (8.24a), is therefore $\psi_{\mathcal{Q}}e^{ikz}$. When the Moon approaches occultation of the radio source, the upper bound on the Fresnel integral begins to diminish from $\sigma_{\max} = +\infty$, and the complex vector on the Cornu spiral begins to oscillate in length (e.g., from A to B in Fig. 8.12b,c) and in phase. The observed flux also oscillates, more and more strongly as geometric

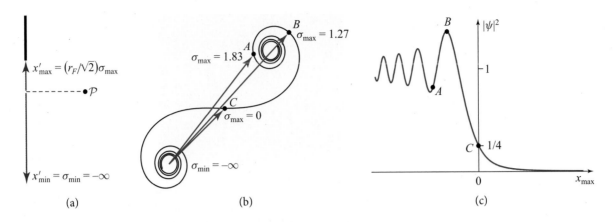

FIGURE 8.12 Diffraction from a straight edge. (a) The straight edge, onto which a plane wave impinges orthogonally, the observation point $\mathcal{P}$, and the vectors that reach to the lower and upper limits of integration, $x'_{min} = -\infty$ and x'_{max} for computation of the diffraction integral (8.23b). (b) The Cornu spiral, showing the arrows that represent the contribution $\Delta\mathcal{E}_\sigma$ to the diffracted field $\psi_{\mathcal{P}}$. (c) The energy-flux diffraction pattern formed by the straight edge. The flux $|\psi|^2 \propto |\Delta\mathcal{E}_\sigma|^2$ is proportional to the squared length of the arrow on the Cornu spiral in panel b.

occultation is approached. At the point of geometric occultation (point C in Fig. 8.12b,c), the complex vector extends from $(-1/2, -1/2)$ to $(0, 0)$, and so the observed wave amplitude is one-half the unocculted value, and the flux is reduced to one-fourth. As the occultation proceeds, the length of the complex vector and the observed flux decrease monotonically to zero, while the phase continues to oscillate.

Historically, diffraction of a radio source's waves by the Moon led to the discovery of quasars—the hyperactive nuclei of distant galaxies. In the early 1960s, a team of Australian and British radio observers led by Cyril Hazard knew that the Moon would occult the powerful radio source 3C273, so they set up their telescope to observe the development of the diffraction pattern as the occultation proceeded. From the pattern's observed times of ingress (passage into the Moon's shadow) and egress (emergence from the Moon's shadow), Hazard determined the coordinates of 3C273 on the sky and did so with remarkable accuracy, thanks to the oscillatory features in the diffraction pattern. These coordinates enabled Maarten Schmidt at the 200-inch telescope on Palomar Mountain to identify 3C273 optically and discover (from its optical redshift) that it was surprisingly distant and consequently had an unprecedented luminosity. It was the first example of a quasar—a previously unknown astrophysical object.

discovery of quasars

In Hazard's occultation measurements, the observing wavelength was $\lambda \sim 0.2$ m. Since the Moon is roughly $z \sim 400{,}000$ km distant, the Fresnel length was about $r_F = \sqrt{\lambda z} \sim 10$ km. The Moon's orbital speed is $v \sim 200$ m s^{-1}, so the diffraction pattern took a time $\sim 5 r_F / v \sim 4$ min to pass through the telescope.

The straight-edge diffraction pattern of Fig. 8.12c occurs universally along the edge of the shadow of any object, so long as the source of light is sufficiently small

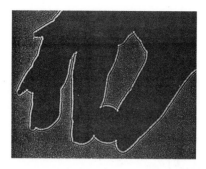

FIGURE 8.13 Fresnel diffraction pattern in the shadow of *Mary's Hand Holding a Dime*—a photograph by Eugene Hecht, from the first figure in Hecht (2017, Chap. 10).

and the shadow's edge bends on lengthscales long compared to the Fresnel length $r_F = \sqrt{\lambda z}$. Examples are the diffraction patterns on the two edges of a slit's shadow in the upper-left curve in Fig. 8.4 (cf. Ex. 8.8) and the diffraction pattern along the edge of a shadow cast by a person's hand in Fig. 8.13.

EXERCISES

Exercise 8.8 *Example: Diffraction Pattern from a Slit*
Derive a formula for the energy-flux diffraction pattern $F(x)$ of a slit with width a, as a function of distance x from the center of the slit, in terms of Fresnel integrals. Plot your formula for various distances z from the slit's plane (i.e., for various values of $r_F/a = \sqrt{\lambda z/a^2}$), and compare with Fig. 8.4.

8.4.4

8.4.4 Circular Apertures: Fresnel Zones and Zone Plates

We have seen how the Fresnel diffraction pattern for a plane wave can be thought of as formed by waves that derive from a patch in the diffracting object's plane a few Fresnel lengths in size. This notion can be made quantitatively useful by reanalyzing an unobstructed wavefront in circular polar coordinates. More specifically, consider a plane wave incident on an aperture $\mathcal{Q}$ that is infinitely large (no obstruction), and define $\varpi \equiv |\mathbf{x}'|/r_F = \sqrt{\frac{1}{2}(\sigma^2 + \tau^2)}$. Then the phase factor in Eq. (8.23b) is $\Delta\varphi = \pi\varpi^2$, so the observed wave coming from the region inside a circle of radius $|\mathbf{x}'| = r_F\varpi$ will be given by

$$\psi_P = -i \int_0^{\varpi} 2\pi\varpi'\, d\varpi'\, e^{i\pi\varpi'^2} \psi_{\mathcal{Q}} e^{ikz}$$

$$= (1 - e^{i\pi\varpi^2})\psi_{\mathcal{Q}} e^{ikz}. \tag{8.25}$$

Now, this integral does not appear to converge as $\varpi \to \infty$. We can see what is happening if we sketch an amplitude-and-phase diagram (Fig. 8.14).

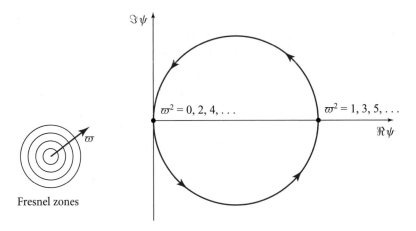

FIGURE 8.14 Amplitude-and-phase diagram for an unobstructed plane wavefront, decomposed into Fresnel zones; Eq. (8.25).

As we integrate outward from $\varpi = 0$, the complex vector has the initial phase retardation of $\pi/2$ but then moves on a semi-circle, so that by the time we have integrated out to a radius of $|\mathbf{x}'| = r_F$ ($\varpi = 1$), the contribution to the observed wave is $\psi_{\mathcal{P}} = 2\psi_{\mathcal{Q}}$ in phase with the incident wave. When the integration has been extended to $\sqrt{2}\, r_F$, ($\varpi = \sqrt{2}$), the circle has been completed, and $\psi_{\mathcal{P}} = 0$! The integral continues on around the same circle over and over again, as the upper-bound radius is further increased; see Fig. 8.14.

Of course, the field must actually have a well-defined value as $\varpi \to \infty$, despite this apparent failure of the integral to converge. To understand how the field becomes well defined, imagine splitting the aperture $\mathcal{Q}$ up into concentric annular rings, known as *Fresnel half-period zones,* with outer radii $\sqrt{n}\, r_F$, where $n = 1, 2, 3 \ldots$. The integral **half-period zones** fails to converge because the contribution from each odd-numbered ring cancels that from an adjacent even-numbered ring. However, the thickness of these rings decreases as $1/\sqrt{n}$, and eventually we must allow for the fact that the incoming wave is not exactly planar; or, equivalently and more usefully, we must allow for the fact that the wave's distant source has some finite angular size. The finite size causes different pieces of the source to have their Fresnel rings centered at slightly different points in the aperture plane, which causes our computation of $\psi_{\mathcal{P}}$ to begin averaging over rings. This averaging forces the tip of the complex vector ($\Re\psi$, $\Im\psi$), where $\Re$ and $\Im$ correspond to the real and imaginary parts, to asymptote to the center of the circle in Fig. 8.14. Correspondingly, due to the averaging, the observed energy flux asymptotes to $|\psi_{\mathcal{Q}}|^2$ [Eq. (8.25) with the exponential $e^{i\pi\varpi^2}$ going to zero].

Although this may not seem to be a particularly wise way to decompose a plane wavefront, it does allow a striking experimental verification of our theory of diffraction. Suppose that we fabricate an aperture (called a *zone plate*) in which, for a **zone plate** chosen wavelength and observation point $\mathcal{P}$ on the optic axis, alternate half-period zones are obscured. Then the wave observed at $\mathcal{P}$ will be the linear sum of several diameters of the circle in Fig. 8.14 and therefore will be far larger than ψ_Q. This strong amplification is confined to our chosen spot on the optic axis; most everywhere else the field's energy flux is reduced, thereby conserving energy. Thus the zone plate behaves like a lens. Because the common area of the half-period zones is $A = n\pi r_F^2 - (n-1)\pi r_F^2 = \pi r_F^2 = \pi\lambda z$, if we construct a zone plate with fixed area A for the zones, its focal length will be $f = z = A/(\pi\lambda)$. For $A =$ (a few square millimeters)—the typical choice—and $\lambda \sim 500$ nm (optical wavelengths), we have $f \sim$ a few meters.

secondary foci

Zone plates are only good lenses when the radiation is monochromatic, since the focal length is wavelength dependent, $f \propto \lambda^{-1}$. They have the further interesting property that they possess secondary foci, where the fields from 3 contiguous zones, or 5 or 7 or so forth, add up coherently (Ex. 8.9).

EXERCISES

Exercise 8.9 *Problem: Zone Plate*

(a) Use an amplitude-and-phase diagram to explain why a zone plate has secondary foci at distances of $f/3$, $f/5$, $f/7$ ....

(b) An opaque, perfectly circular disk of diameter D is placed perpendicular to an incoming plane wave. Show that, at distances r such that $r_F \ll D$, the disk casts a rather sharp shadow, but at the precise center of the shadow there must be a bright spot.[5] How bright?

Exercise 8.10 *Problem: Spy Satellites*

Telescopes can also look down through the same atmospheric irregularities as those mentioned in Sec. 8.3.2 (see also Box 9.2). In what important respects will the optics differ from those for ground-based telescopes looking upward?

8.5

8.5 Paraxial Fourier Optics

We have developed a linear theory of wave optics that has allowed us to calculate diffraction patterns in the Fraunhofer and Fresnel limiting regions. That these calculations agree with laboratory measurements provides some vindication of the theory and the assumptions implicit in it. We now turn to practical applications of these ideas, specifically, to the acquisition and processing of images by instruments operating throughout the electromagnetic spectrum. As we shall see, these instruments rely

---

5. Siméon Poisson predicted the existence of this spot as a consequence of Fresnel's wave theory of light, in order to demonstrate that Fresnel's theory was wrong. However, François Arago (who was briefly, in 1848, premier of France) quickly demonstrated experimentally that the bright spot existed. It is now called the Poisson spot (despite Poisson's skepticism) or the Arago spot.

on an extension of paraxial geometric optics (Sec. 7.4) to situations where diffraction effects are important. Because of the central role played by Fourier transforms in diffraction [e.g., Eq. (8.11a)], the theory underlying these instruments is called *paraxial Fourier optics,* or just *Fourier optics.*

Although the conceptual framework and mathematical machinery for image processing by Fourier optics were developed in the nineteenth century, Fourier optics was not widely exploited until the second half of the twentieth century. Its maturation was driven in part by a growing recognition of similarities between optics and communication theory—for example, the realization that a microscope is simply an *image-processing system.* The development of electronic computation has also triggered enormous strides; computers are now seen as extensions of optical devices, and vice versa. It is a matter of convenience, accuracy, economics, and practicality to decide which parts of the image processing are carried out with mirrors, lenses, and the like, and which parts are performed numerically.

image processing with mirrors, lenses, etc. plus computing

One conceptually simple example of optical image processing is an improvement in one's ability to identify a faint star in the Fraunhofer diffraction rings ("fringes") of a much brighter star. As we shall see [Eq. (8.31) and subsequent discussion], the bright image of a source in a telescope's or microscope's focal plane has the same Airy diffraction pattern as we met in Eq. (8.18) and Fig. 8.8. If the shape of that image could be changed from the ring-endowed Airy pattern to a Gaussian, then it would be far easier to identify a nearby feature or faint star. One way to achieve this would be to attenuate the incident radiation at the telescope aperture in such a way that, immediately after passing through the aperture, it has a Gaussian profile instead of a sharp-edged profile. Its Fourier transform (the diffraction pattern in the focal plane) would then also be Gaussian. Such a Gaussian-shaped attenuation is difficult to achieve in practice, but it turns out—as we shall see—that there are easier options.

Gaussian aperture

Before exploring these options, we must lay some foundations, beginning with the concept of coherent illumination in Sec. 8.5.1, and then moving on to point-spread functions in Sec. 8.5.2.

8.5.1 Coherent Illumination

8.5.1

If the radiation arriving at the input of an optical system derives from a single source (e.g., a point source that has been collimated into a parallel beam by a converging lens), then the radiation is best described by its complex amplitude ψ (as we are doing in this chapter). An example might be a biological specimen on a microscope slide, illuminated by an external point source, for which the phases of the waves leaving different parts of the slide are strongly correlated with one another. This is called *coherent illumination.* If, by contrast, the source is self-luminous and of nonnegligible size, with the atoms or molecules in its different parts radiating independently—for example, a cluster of stars—then the phases of the radiation from different parts are uncorrelated, and it may be the radiation's energy flux, not its complex amplitude, that obeys well-defined (nonprobabilistic) evolution laws. This is called *incoherent illumination.* In this chapter we develop Fourier optics for a coherently illuminating source (the kind of illumination tacitly assumed in previous sections of the chapter). A parallel theory with a similar vocabulary can be developed for incoherent illumination, and some of the foundations for it are laid in Chap. 9. In Chap. 9, we also develop a more precise formulation of the concept of *coherence.*

8.5.2 Point-Spread Functions

In our treatment of paraxial geometric optics (Sec. 7.4), we showed how it is possible to regard a group of optical elements as a sequence of linear devices and relate the output rays to the input by linear operators (i.e., matrices). This chapter's theory of diffraction is also linear, and so a similar approach can be followed. As in Sec. 7.4, we restrict attention to small angles relative to some optic axis ("paraxial Fourier optics"). We describe the wave field at some distance z_j along the optic axis by the function $\psi_j(\mathbf{x})$, where $\mathbf{x}$ is a 2-dimensional vector perpendicular to the optic axis (as in Fig. 8.5). If we consider a single linear optical device, then we can relate the output field ψ_2 at z_2 to the input ψ_1 at z_1 using a Green's function denoted $P_{21}(\mathbf{x}_2, \mathbf{x}_1)$:[6]

$$\psi_2(\mathbf{x}_2) = \int P_{21}(\mathbf{x}_2, \mathbf{x}_1) d\Sigma_1 \psi_1. \tag{8.26}$$

If ψ_1 were a delta function, then the output would be simply given by the function P_{21}, up to normalization. For this reason, P_{21} is usually known as the *point-spread function.* Alternatively, we can think of it as a *propagator.* If we now combine two optical devices sequentially, so the output ψ_2 of the first device is the input to the second, then the point-spread functions combine in the natural manner of any linear propagator to give a total point-spread function

$$P_{31}(\mathbf{x}_3, \mathbf{x}_1) = \int P_{32}(\mathbf{x}_3, \mathbf{x}_2) d\Sigma_2 P_{21}(\mathbf{x}_2, \mathbf{x}_1). \tag{8.27}$$

Just as the simplest matrix for paraxial, geometric-optics propagation is that for free propagation through some distance d, so also the simplest point-spread function is that for free propagation. From Eq. (8.22) we see that it is given by

$$P_{21} = \frac{-ik}{2\pi d} e^{ikd} \exp\left(\frac{ik(\mathbf{x}_2 - \mathbf{x}_1)^2}{2d}\right)$$

for free propagation through a distance $d = z_2 - z_1$. \qquad (8.28)

Green's function, point-spread function, propagator

combining point-spread functions

free-propagation point-spread function

---

6. The approach followed here has an interesting history. It originated in the treatment of Markov processes by Norbert Wiener and others (see Sec 6.3). It was then cleverly adopted for use in quantum mechanics and quantum field theory by Paul Dirac, Richard Feynman, and others, where it constitutes the path integral formulation. This, in turn, suggested the simpler adaptation to problems in classical optics laid out here.

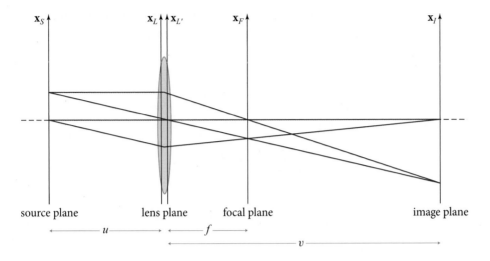

FIGURE 8.15 Wave theory of a single converging lens. The focal plane is a distance f (lens focal length) from the lens plane; and the image plane is a distance $v = fu/(u - f)$ from the lens plane.

Note that this P_{21} depends only on $\mathbf{x}_2 - \mathbf{x}_1$ as it should, and not on $\mathbf{x}_2$ or $\mathbf{x}_1$ individually, because there is joint translational invariance in the transverse $\mathbf{x}_1$, $\mathbf{x}_2$ planes.

A thin lens adds or subtracts an extra phase $\Delta\varphi$ to the wave, and $\Delta\varphi$ depends quadratically on distance $|\mathbf{x}|$ from the optic axis, so the angle of deflection, which is proportional to the gradient of the extra phase, will depend linearly on $\mathbf{x}$. Correspondingly, the point-spread function for a thin lens with focal length f is

$$P_{21} = \exp\left(\frac{-ik|\mathbf{x}_1|^2}{2f}\right) \delta(\mathbf{x}_2 - \mathbf{x}_1). \tag{8.29}$$

thin-lens point-spread function

For a converging lens, f is positive; for a diverging one, it is negative.

8.5.3 Abbé's Description of Image Formation by a Thin Lens

8.5.3

We can use the two point-spread functions (8.28) and (8.29) to give a wave description of the production of images by a single converging lens, in parallel to the geometric-optics description of Fig. 7.6. This description was formulated by Ernst Abbé in 1873. We develop Abbé's description in two stages. First, we propagate the wave from the source plane S a distance u in front of the lens, through the lens L, and to its focal plane F a distance f behind the lens (Fig. 8.15). Then we propagate the wave a further distance $v - f$ from the focal plane to the image plane. We know from geometric optics that $v = fu/(u - f)$ [Eq. (7.58)]. We restrict ourselves to $u > f$, so v is positive and the lens forms a real, inverted image.

Using Eqs. (8.27)–(8.29), we obtain the following expression for the propagator from the source plane S through the lens plane L to the focal plane F:

$$
\begin{aligned}
P_{FS} &= \int P_{FL'} d\Sigma_{L'} P_{L'L} d\Sigma_L P_{LS} \\
&= \int \frac{-ik}{2\pi f} e^{ikf} \exp\left(\frac{ik(\mathbf{x}_F - \mathbf{x}'_L)^2}{2f}\right) d\Sigma_{L'} \delta(\mathbf{x}_{L'} - \mathbf{x}_L) \exp\left(\frac{-ik|\mathbf{x}_L|^2}{2f}\right) \\
&\quad \times d\Sigma_L \frac{-ik}{2\pi u} e^{iku} \exp\left(\frac{ik(\mathbf{x}_L - \mathbf{x}_S)^2}{2u}\right) \\
&= \frac{-ik}{2\pi f} e^{ik(f+u)} \exp\left(\frac{-ikx_F^2}{2(v-f)}\right) \exp\left(\frac{-ik\mathbf{x}_F \cdot \mathbf{x}_S}{f}\right).
\end{aligned}
\tag{8.30}
$$

Here we have extended all integrations to $\pm\infty$ and have used the values of the Fresnel integrals at infinity, $\mathcal{E}(\pm\infty) = \pm(1+i)/2$, to get the expression on the last line. The wave in the focal plane is given by $\psi_F(\mathbf{x}_F) = \int P_{FS} d\Sigma_S \psi_S(\mathbf{x}_S)$, which integrates to

$$
\boxed{\psi_F(\mathbf{x}_F) = -\frac{ik}{2\pi f} e^{ik(f+u)} \exp\left(\frac{-ikx_F^2}{2(v-f)}\right) \tilde{\psi}_S(\mathbf{x}_F/f).}
\tag{8.31}
$$

Here

$$
\tilde{\psi}_S(\boldsymbol{\theta}) = \int d\Sigma_S \psi_S(\mathbf{x}_S) e^{-ik\boldsymbol{\theta} \cdot \mathbf{x}_S}.
\tag{8.32}
$$

Equation (8.31) states that the field in the focal plane is, apart from a phase factor, proportional to the Fourier transform of the field in the source plane [recall our optics convention Eq. (8.11a) for the Fourier transform]. In other words, *the focal-plane field is the Fraunhofer diffraction pattern of the input wave.* That this has to be the case can be understood from Fig. 8.15. The focal plane F is where the converging lens brings parallel rays from the source plane to a focus. By doing so, the lens in effect brings in from "infinity" the Fraunhofer diffraction pattern of the source [Eq. (8.11a)][7] and places it into the focal plane.

It now remains to propagate the final distance from the focal plane to the image plane. We do so with the free-propagation point-spread function (8.28): $\psi_I = \int P_{IF} d\Sigma_F \psi_F$, which integrates to

$$
\boxed{\psi_I(\mathbf{x}_I) = -\left(\frac{u}{v}\right) e^{ik(u+v)} \exp\left(\frac{ikx_I^2}{2(v-f)}\right) \psi_S(\mathbf{x}_S = -\mathbf{x}_I u/v).}
\tag{8.33}
$$

---

7. In Eq. (8.11a), the input wave at the system's entrance aperture is $\psi_S = \psi' t(\mathbf{x}) \propto t(\mathbf{x})$ [Eq. (8.6) and Fig. 8.2], the Fraunhofer diffraction pattern is $\psi_{\mathcal{P}} \propto \tilde{t}(\boldsymbol{\theta})$, and the lens produces the focal-plane field $\tilde{\psi}_F \propto \tilde{t}(\mathbf{x}_F/f)$.

Thus (again ignoring a phase factor) the wave in the image plane is just a magnified and inverted version of the wave in the source plane, as we might have expected from geometric optics. In words, the lens acts by taking the Fourier transform of the source and then takes the Fourier transform again to recover the source structure.

8.5.4 Image Processing by a Spatial Filter in the Focal Plane of a Lens: High-Pass, Low-Pass, and Notch Filters; Phase-Contrast Microscopy

The focal plane of a lens is a convenient place to process an image by altering its Fourier transform. This process, known as *spatial filtering*, is a powerful technique. We shall gain insight into its power via several examples.

In each of these examples, we assume for simplicity the one-lens system of Fig. 8.15, for which we worked out Abbé's description in the previous section. If the source wave has planar phase fronts parallel to the source plane so ψ_S is real, then the output wave in the image plane has spherical phase fronts, embodied in the phase factor $\exp[ikx_i^2/(2(v-f))]$. If, instead, one wants the output wave to have the same planar phase fronts as the input wave, one can achieve that by using a two-lens system with the lenses separated by the sum of the lenses' focal lengths and altering the Fourier transform occurring in the common focal plane between them (Ex. 8.11).

LOW-PASS FILTER: CLEANING A LASER BEAM

In low-pass filtering, a small circular aperture or "stop" is introduced into the focal plane, thereby allowing only the low-order spatial Fourier components (long-wavelength components) to be transmitted to the image plane. This produces a considerable smoothing of the wave. An application is to the output beam from a laser (Chap. 10), which ought to be smooth but in practice has high spatial frequency structure (high transverse wave numbers, short wavelengths) on account of noise and imperfections in the optics. A low-pass filter can be used to clean the beam. In the language of Fourier transforms, if we multiply the transform of the source, in the focal plane, by a small-diameter circular aperture function, we will thereby convolve the image with a broad Airy-disk smoothing function.

HIGH-PASS FILTER: ACCENTUATING AN IMAGE'S FEATURES

Conversely, we can exclude the low spatial frequencies with a high-pass filter (e.g., by placing an opaque circular disk in the focal plane, centered on the optic axis). This accentuates boundaries and discontinuities in the source's image and can be used to highlight features where the gradient of the brightness is large.

NOTCH FILTER: REMOVING PIXELATION FROM AN IMAGE

Another type of filter is used when the image is pixelated and thus has unwanted structure with wavelength equal to pixel size. A narrow range of frequencies centered around this spatial frequency is removed by putting an appropriate filter in the focal plane; this is sometimes called a "notch filter."

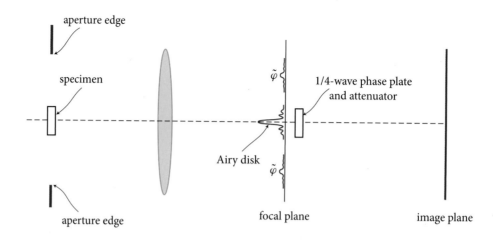

FIGURE 8.16 Schematic phase-contrast microscope.

PHASE-CONTRAST MICROSCOPY

Phase-contrast microscopy (Fig. 8.16) is a useful technique for studying small objects, such as transparent biological specimens, that modify the phase of coherent illuminating light but not its amplitude. Suppose that the differential phase change in the object is small, $|\varphi| \ll 1$, as often is the case for biological specimens. We can then write the field just after it passes through the specimen as

$$\psi_S(\mathbf{x}) = H(\mathbf{x})e^{i\varphi(\mathbf{x})} \simeq H(\mathbf{x}) + i\varphi(\mathbf{x})H(\mathbf{x}). \tag{8.34}$$

Here H is the microscope's aperture function, unity for $|\mathbf{x}| < D/2$ and zero for $|\mathbf{x}| > D/2$, with D the aperture diameter. The energy flux is not modulated, and therefore the effect of the specimen on the wave is hard to observe unless one is clever.

Equation (8.34) and the linearity of the Fourier transform imply that the wave in the focal plane is the sum of (i) the Fourier transform of the aperture function [Eq. (8.18) and Ex. 8.4] and (ii) the transform of the phase function convolved with that of the aperture [in which the fine-scale variations of the phase function dominate and push $\tilde{\varphi}$ to large radii in the focal plane (Fig. 8.16), where the aperture has little influence]:

$$\psi_F \sim \text{jinc}\left(\frac{kD|\mathbf{x}_F|}{2f}\right) + i\tilde{\varphi}\left(\frac{\mathbf{x}_F}{f}\right), \quad \text{where} \quad \text{jinc}(x) = J_1(x)/x. \tag{8.35}$$

If a high-pass filter is used to remove the Airy disk completely, then the remaining wave field in the image plane will be essentially φ magnified by v/u. However, the energy flux $F \propto (\varphi v/u)^2$ will be quadratic in the phase, and so the contrast in the image will still be small. A better technique is to phase shift the Airy disk in the focal plane by $\pi/2$ so that the two terms in Eq. (8.35) are in phase. The flux variations, $F \sim (1 \pm \varphi)^2 \simeq 1 \pm 2\varphi$, will now be linear in the phase φ. An even better procedure is

converting phase changes
to amplitude changes

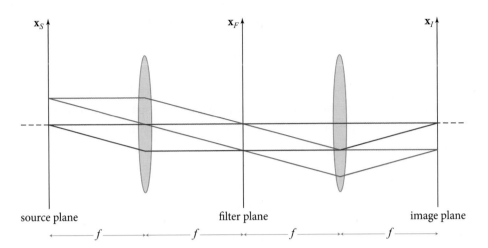

x_S x_F x_I

source plane   filter plane   image plane

$\longleftarrow f \longrightarrow \longleftarrow f \longrightarrow \longleftarrow f \longrightarrow \longleftarrow f \longrightarrow$

FIGURE 8.17 Two-lens system for spatial filtering.

to attenuate the Airy disk until its amplitude is comparable with the rms value of φ and also phase shift it by $\pi/2$ (as indicated by the "1/4-wave phase plate and attenuator" in Fig. 8.16). This will maximize the contrast in the final image. Analogous techniques are used in communications to interconvert amplitude-modulated and phase-modulated signals.

Exercise 8.11 **Example: Two-Lens Spatial Filter*

Figure 8.17 depicts a two-lens system for spatial filtering (also sometimes called a "$4f$ system," since it involves five special planes separated by four intervals with lengths equal to the common focal length f of the two lenses). Develop a description, patterned after Abbé's, of image formation with this system. Most importantly, do the following.

(a) Show that the field at the filter plane is

$$\psi_F(\mathbf{x}_F) = -\frac{ik}{2\pi f} e^{2ikf} \tilde{\psi}_S(\mathbf{x}_F/f). \tag{8.36a}$$

This is like Eq. (8.31) for the one-lens system but with the spatially dependent phase factor $\exp[-ikx_F^2/(2(v-f))]$ gone, so aside from a multiplicative constant, the filter-plane field is precisely the Fourier transform of the source-plane field.

(b) In the filter plane we place a filter whose transmissivity we denote $\tilde{K}(-\mathbf{x}_F/f)$, so it is (proportional to) the filter-plane field that would be obtained from some source field $K(-\mathbf{x}_S)$. Using the optics conventions (8.11) for the Fourier transform and its inverse, show that the image-plane field, with the filter present, is

$$\psi_I(\mathbf{x}_I) = -e^{4ikf} \Psi_S(\mathbf{x}_S = -\mathbf{x}_I). \tag{8.36b}$$

Here Ψ_S is the convolution of the source field and the filter function

$$\Psi_S(\mathbf{x}_S) = \int K(\mathbf{x}_S - \mathbf{x}')\psi_S(\mathbf{x}')d\Sigma'. \qquad (8.36c)$$

In the absence of the filter, Ψ_S is equal to ψ_S, so Eq. (8.36b) is like the image-plane field (8.33) for the single-lens system, but with the spatially dependent phase factor $\exp[-ikx_F^2/(2(v-f))]$ gone. Thus ψ_I is precisely the same as the inverted ψ_S, aside from an overall phase.

Exercise 8.12 *Problem: Convolution via Fourier Optics*

(a) Suppose that you have two thin sheets with transmission functions $t = g(x, y)$ and $t = h(x, y)$, and you wish to compute via Fourier optics the convolution

$$g \otimes h(x_o, y_o) \equiv \int \int g(x, y)h(x_o - x, y_o - y)\, dx\, dy. \qquad (8.37)$$

Devise a method for doing so using Fourier optics. [Hint: Use several lenses and a projection screen with a pinhole through which passes light whose energy flux is proportional to the convolution; place the two sheets at strategically chosen locations along the optic axis, and displace one of the two sheets transversely with respect to the other.]

(b) Suppose you wish to convolve a large number of different pairs of 1-dimensional functions $\{g_1h_1\}$, $\{g_2h_2\}$, ... simultaneously; that is, you want to compute

$$g_j \otimes h_j(x_o) \equiv \int g_j(x)h_j(x_o - x)dx \qquad (8.38)$$

for $j = 1, 2, \ldots$. Devise a way to do this via Fourier optics using appropriately constructed transmissive sheets and cylindrical lenses.

Exercise 8.13 ***Example: Transmission Electron Microscope*

In a transmission electron microscope, electrons, behaving as waves, are prepared in near-plane-wave quantum wave-packet states with transverse sizes large compared to the object ("sample") being imaged. The sample, placed in the source plane of Fig. 8.15, is sufficiently thin—with a transmission function $t(\mathbf{x})$—that the electron waves can pass through it, being diffracted in the process. The diffracted waves travel through the lens shown in Fig. 8.15 (the *objective lens*), then onward to and through a second lens, called the *projector lens,* which focuses them onto a fluorescent screen that shines with an energy flux proportional to the arriving electron flux. At least two lenses are needed, as in the simplest of optical microscopes and telescopes (Figs. 7.7 and 7.8), and for the same reason: to make images far larger than could be achieved with a single lens.

(a) The electrons all have the same kinetic energy $E \sim 200$ keV, to within an energy spread $\Delta E \sim 1$ eV. What is the wavelength λ of their nearly plane-wave quantum wave functions, and what is their fractional wavelength spread $\Delta\lambda/\lambda$? Your answer for λ (∼a few picometers) is so small that electron microscopy can be used to study atoms and molecules. Contrast this with light's million-fold longer wavelength ∼1 μm, which constrains it to imaging objects a million times larger than atoms.

(b) Explain why the paraxial Fourier-optics formalism that we developed in Sec. 8.5 can be used without change to analyze this electron microscope. This is true even though the photons of an ordinary microscope are in states with mean occupation numbers η huge compared to unity while the electrons have $\eta \ll 1$, and the photons have zero rest mass while the electrons have finite rest mass m with roughly the same magnitude as their kinetic energies, $E \sim mc^2$.

(c) Suppose that each magnetic lens in the electron microscope is made of two transverse magnetic quadrupoles, as described in Sec. 7.4.2. Show that, although these quadrupoles are far from axisymmetric, their combined influence on each electron's wave function is given by the axisymmetric thin-lens point-spread function (8.29) with focal length (7.66), to within the accuracy of the analysis of Sec. 7.4.2. [In practice, higher-order corrections make the combined lens sufficiently nonaxisymmetric that electron microscopes do not use this type of magnetic lens. Instead they use a truly axisymmetric lens in which the magnetic field lines lie in planes of constant azimuthal angle ϕ, and the field lines first bend the electron trajectories into helices (give them motion in the ϕ direction), then bend them radially, and then undo the helical motion.]

(d) By appropriate placement of the projector lens, the microscope can produce, in the fluorescing plane, either a vastly enlarged image of the source-plane sample, $|t(\mathbf{x})|^2$, or a large image of the modulus of that object's Fourier transform (the object's diffraction pattern), $|\tilde{t}(\mathbf{k})|^2$. Explain how each of these is achieved.

8.5.5 Gaussian Beams: Optical Cavities and Interferometric Gravitational-Wave Detectors

The mathematical techniques of Fourier optics enable us to analyze the structure and propagation of light beams that have Gaussian profiles. Such Gaussian beams are the natural output of ideal lasers; they are the real output of spatially filtered lasers; and they are widely used for optical communication, interferometry, and other practical applications. Moreover, they are the closest one can come in the real world of wave optics to the idealization of a geometric-optics pencil beam.

Consider a beam that is precisely plane-fronted, with a Gaussian profile, at location $z = 0$ on the optic axis:

$$\psi_0 = \exp\left(\frac{-\varpi^2}{\sigma_0^2}\right); \qquad (8.39)$$

here $\varpi = |\mathbf{x}|$ is radial distance from the optic axis. The form of this same wave at a distance z farther down the optic axis can be computed by propagating this ψ_0 using the point-spread function (8.28) (with the distance d replaced by z). The result is

freely propagating Gaussian beam

$$\psi_z = \frac{\sigma_0}{\sigma_z} \exp\left(\frac{-\varpi^2}{\sigma_z^2}\right) \exp\left[i\left(\frac{k\varpi^2}{2R_z} + kz - \arctan\frac{z}{z_0}\right)\right],$$

(8.40a)

where

$$z_0 = \frac{k\sigma_0^2}{2} = \frac{\pi\sigma_0^2}{\lambda}, \quad \sigma_z = \sigma_0(1 + z^2/z_0^2)^{1/2}, \quad R_z = z(1 + z_0^2/z^2),$$

(8.40b)

and a subscript z indicates that this quantity is a function of z. These equations for the freely propagating Gaussian beam are valid for negative z as well as positive.

From these equations we learn the following properties of the beam:

- The beam's cross sectional energy-flux distribution

$$F \propto |\psi_z|^2 \propto \exp(-\varpi^2/\sigma_z^2)$$

remains a Gaussian as the wave propagates, with a beam radius

beam radius

$$\sigma_z = \sigma_0\sqrt{1 + z^2/z_0^2}$$

that is a minimum at $z = 0$ (the beam's *waist*) and grows away from the waist, both forward and backward, in just the manner to be expected from our uncertainty-principle discussion of wavefront spreading [Eq. (8.9)]. At distances $|z| \ll z_0$ from the waist location (corresponding to a Fresnel length $r_F = \sqrt{\lambda|z|} \ll \sqrt{\pi}\sigma_0$), the beam radius is nearly constant; this is the Fresnel region. At distances $z \gg z_0$ ($r_F \gg \sqrt{\pi}\sigma_0$), the beam radius increases linearly [i.e., the beam spreads with an opening angle $\theta = \sigma_0/z_0 = \lambda/(\pi\sigma_0)$]; this is the Fraunhofer region.

- The beam's wavefronts (surfaces of constant phase) have phase

$$\varphi = k\varpi^2/(2R_z) + kz - \arctan(z/z_0) = \text{constant}.$$

Gouy phase

The arctan term (called the wave's *Gouy phase*) varies far far more slowly with changing z than does the kz term, so the wavefronts are almost precisely $z = -\varpi^2/2R_z + \text{const}$, which is a segment of a sphere of radius R_z. Thus, the wavefronts are spherical, with radii of curvature $R_z = z(1 + z_0^2/z^2)$, which is infinite (flat phase fronts) at the waist $z = 0$, decreases to a minimum of $2z_0$ at $z = z_0$ (boundary between Fresnel and Fraunhofer regions and beginning of substantial wavefront spreading), and then increases as z (gradual flattening of spreading wavefronts) when one moves deep into the Fraunhofer region.

- The Gaussian beam's form [Eqs. (8.40)] at some arbitrary location is fully characterized by three parameters: the wavelength $\lambda = 2\pi/k$, the distance z to the waist, and the beam radius σ_0 at the waist [from which one can compute the local beam radius σ_z and the local wavefront radius of curvature R_z via Eqs. (8.40b)].

One can easily compute the effects of a thin lens on a Gaussian beam by folding the ψ_z at the lens's location into the lens point-spread function (8.29). The result is a phase change that preserves the general Gaussian form of the wave but alters the distance z to the waist and the radius σ_0 at the waist. Thus, by judicious placement of lenses or mirrors and with judicious choices of the lenses' and mirrors' focal lengths, one can tailor the parameters of a Gaussian beam to fit whatever optical device one is working with. For example, if one wants to send a Gaussian beam into a self-focusing optical fiber (Exs. 7.8 and 8.14), one should place its waist at the entrance to the fiber and adjust its waist size there to coincide with that of the fiber's Gaussian mode of propagation (the mode analyzed in Ex. 8.14). The beam will then enter the fiber smoothly and will propagate steadily along the fiber, with the effects of the transversely varying index of refraction continually compensating for the effects of diffraction so as to keep the phase fronts flat and the waist size constant.

tailoring the beam parameters

Gaussian beams are used (among many other places) in interferometric gravitational-wave detectors, such as LIGO (the Laser Interferometer Gravitational-Wave Observatory). We shall learn how these *GW interferometers* work in Sec. 9.5. For the present, all we need to know is that a GW interferometer entails an optical cavity formed by mirrors facing each other, as in Fig. 7.9. A Gaussian beam travels back and forth between the two mirrors, its light superposing on itself coherently after each round trip—the light *resonates* in the cavity formed by the two mirrors. Each mirror hangs from an overhead support, and when a gravitational wave passes, it pushes the hanging mirrors back and forth with respect to each other, causing the cavity to lengthen and shorten by a tiny fraction of a light wavelength. This puts a minuscule phase shift on the resonating light, which is measured by allowing some of the light to leak out of the cavity and interfere with light from another, similar cavity (see Sec. 9.5).

use of Gaussian beams in LIGO

For the light to resonate in the cavity, the mirrors' surfaces must coincide with the Gaussian beam's wavefronts. Suppose that the mirrors are identical, with radii of curvature R, and are separated by a distance $L = 4$ km, as in LIGO. Then the beam must be symmetric around the center of the cavity, so its waist must be halfway between the mirrors. What is the smallest that the beam radius can be at the mirrors' locations $z = \pm L/2 = \pm 2$ km? From $\sigma_z = \sigma_0(1 + z^2/z_0^2)^{1/2}$ together with $z_0 = \pi\sigma_0^2/\lambda$, we see that $\sigma_{L/2}$ is minimized when $z_0 = L/2 = 2$ km. If the wavelength is $\lambda = 1.064\ \mu$m (Nd:YAG—neodymium-doped yttrium aluminum garnet—laser light, as in LIGO), then the beam radii at the waist and at the mirrors are $\sigma_0 = \sqrt{\lambda z_0/\pi} = \sqrt{\lambda L/2\pi} = 2.6$ cm, and $\sigma_{L/2} = \sqrt{2}\sigma_0 = 3.7$ cm, respectively, and the mirrors' radii of curvature are $R_{L/2} = L = 4$ km. This was approximately the regime of parameters used for

LIGO's initial GW interferometers, which carried out a 2-year-long search for gravitational waves from autumn 2005 to autumn 2007 and then, after some sensitivity improvements, a second long search in 2009 and 2010.

advanced LIGO

A new generation of GW interferometers, called "Advanced LIGO" (LIGO Scientific Collaboration, 2015), began operating in summer 2015 and shortly thereafter discovered gravitational waves. In these GW interferometers, the spot sizes on the mirrors are enlarged, so as to reduce thermal noise by averaging over a much larger spatial sampling of thermal fluctuations of the mirror surfaces (cf. Secs. 6.8.2 and 11.9.2). How were the spot sizes enlarged? From Eqs. (8.40b) we see that, in the limit $z_0 = \pi \sigma_0^2/\lambda \to 0$, the mirrors' radii of curvature approach the cavity half-length, $R_{L/2} \to L/2$, and the beam radii on the mirrors diverge as $\sigma_{L/2} \to L\lambda/(2\pi\sigma_0) \to \infty$. This is the same instability as we discovered in the geometric-optics limit (Ex. 7.12). Advanced LIGO takes advantage of this instability by moving toward the near-unstable regime, causing the beams on the mirrors to enlarge. In Advanced LIGO's semifinal design, the mirrors' radii of curvature were set at $R_{L/2} = 2.076$ km, just 4% above the unstable point $R = L/2 = 2$ km; and Eqs. (8.40b) then tell us that $\sigma_0 = 1.15$ cm, $z_0 = 0.389$ km $\ll L/2 = 2$ km, and σ_z was pushed up by nearly a factor of two, to $\sigma_z = 6.01$ cm. The mirrors are deep into the Fraunhofer, wavefront-spreading region. In the final design the optical cavity was made slightly asymmetric with mirror radii of curvature a bit below and a bit above 2.076 km (Table 1 of LIGO Scientific Collaboration, 2015).

EXERCISES

Exercise 8.14 *Problem: Guided Gaussian Beams*
Consider a self-focusing optical fiber discussed in Ex. 7.8, in which the refractive index is

$$n(\mathbf{x}) = n_o(1 - \alpha^2 r^2)^{1/2}, \tag{8.41}$$

where $r = |\mathbf{x}|$.

(a) Write down the Helmholtz equation in cylindrical polar coordinates and seek an axisymmetric mode for which $\psi = R(r)Z(z)$, where R and Z are functions to be determined, and z measures distance along the fiber. In particular, show that there exist modes with a Gaussian radial profile that propagate along the fiber without spreading.

(b) Compute the group and phase velocities along the fiber for this mode.

Exercise 8.15 *Example: Noise Due to Scattered Light in LIGO*
In LIGO and other GW interferometers, one potential source of noise is scattered light. When the Gaussian beam in one of LIGO's cavities reflects off a mirror, a small portion of the light gets scattered toward the walls of the cavity's vacuum tube. Some of this scattered light can reflect or scatter off the tube wall and then propagate toward the distant mirror, where it scatters back into the Gaussian beam; see Fig. 8.18a (without

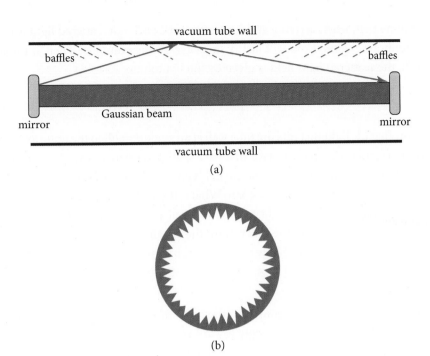

FIGURE 8.18 (a) Scattered light in LIGO's beam tube. (b) Cross section of a baffle used to reduce the noise due to scattered light.

the baffles, which are shown dashed). This is troublesome because the tube wall vibrates from sound-wave excitations and seismic excitations, and those vibrations put a phase shift on the scattered light. Although the fraction of all the light that scatters in this way is tiny, the phase shift is huge compared to that produced in the Gaussian beam by gravitational waves. When the tiny amount of scattered light with its huge oscillating phase shift recombines into the Gaussian beam, it produces a net Gaussian-beam phase shift that can be large enough to mask a gravitational wave. This exercise explores some aspects of this scattered-light noise and its control.

(a) The scattering of Gaussian-beam light off the mirror is caused by bumps in the mirror surface (imperfections). Denote by $h(\mathbf{x})$ the height of the mirror surface relative to the desired shape (a segment of a sphere with radius of curvature that matches the Gaussian beam's wavefronts). Show that, if the Gaussian-beam field emerging from a perfect mirror is $\psi^G(\mathbf{x})$ [Eq. (8.40a)] at the mirror plane, then the beam emerging from the actual mirror is $\psi'(\mathbf{x}) = \psi^G(\mathbf{x}) \exp[-i2kh(\mathbf{x})]$. The magnitude of the mirror irregularities is very small compared to a wavelength, so $|2kh| \ll 1$, and the wave field emerging from the mirror is $\psi'(\mathbf{x}) = \psi^G(\mathbf{x})[1 - i2kh(\mathbf{x})]$. Explain why the factor 1 does not contribute at all to the scattered light (where does its light go?), so the scattered light field emerging from the mirror is

$$\psi^S(\mathbf{x}) = -i\psi^G(\mathbf{x})2kh(\mathbf{x}).\qquad(8.42)$$

(b) Assume that, when arriving at the vacuum-tube wall, the scattered light is in the Fraunhofer region. (You will justify this later in this example.) Then at the tube wall, the scattered light field is given by the Fraunhofer formula

$$\psi^S(\boldsymbol{\theta}) \propto \int \psi^G(\mathbf{x}) k h(\mathbf{x}) e^{ik\mathbf{x}\cdot\boldsymbol{\theta}} d\Sigma. \tag{8.43}$$

Show that the light that hits the tube wall at an angle $\theta = |\boldsymbol{\theta}|$ to the optic axis arises from irregularities in the mirror that have spatial wavelengths $\lambda_{\text{mirror}} \sim \lambda/\theta$. The radius of the beam tube is $\mathcal{R} = 60$ cm in LIGO, and the length of the tube (distance between cavity mirrors) is $L = 4$ km. What is the spatial wavelength of the mirror irregularities that scatter light to the tube wall at distances $z \sim L/2$ (which can then reflect or scatter off the wall toward the distant mirror and there scatter back into the Gaussian beam)? Show that for these irregularities, the tube wall is indeed in the Fraunhofer region. [Hint: The irregularities have a coherence length of only a few wavelengths λ_{mirror}.]

(c) In the initial LIGO interferometers, the mirrors' scattered light consisted of two components: one peaked strongly toward small angles so it hit the distant tube wall (e.g., at $z \sim L/2$), and the other roughly isotropically distributed. What is the size of the irregularities that produced the isotropic component?

(d) To reduce substantially the amount of scattered light reaching the distant mirror via reflection or scattering from the tube wall, a set of baffles was installed in the tube, in such a way as to hide the wall from scattered light (dashed lines in Fig. 8.18). The baffles have an angle of 35° to the tube wall, so when light hits a baffle, it reflects at a steep angle, $\sim$70° toward the opposite tube wall, and after a few bounces gets absorbed. However, a small portion of the scattered light can now *diffract* off the top of each baffle and then propagate to the distant mirror and scatter back into the main beam. Especially troublesome is the case of a mirror in the center of the beam tube's cross section, because light that scatters off such a mirror travels nearly the same total distance from the mirror to the top of some baffle and then to the distant mirror, independent of the azimuthal angle ϕ on the baffle at which it diffracts. There is then a danger of *coherent superposition* of all scattered light that diffracts off all angular locations around any given baffle—and coherent superposition means a much enlarged net noise (a variant of the Poisson or Arago spot discussed in the footnote to Ex. 8.9). To protect against any such coherence, the baffles in the LIGO beam tubes are serrated (i.e., they have saw-tooth edges), and the heights of the teeth are drawn from a random (Gaussian) probability distribution (Fig. 8.18b). The typical tooth heights are large enough to extend through about six Fresnel zones. How wide is each Fresnel zone at the baffle location, and correspondingly, how high must be the typical baffle tooth? By approximately how much do the random serrations reduce the light-scattering noise, relative to what it would be with no serrations and with coherent scattering? [Hint: See part e. There are two ways that the noise is reduced: (i) The breaking of coherence of scattered light by the randomness of serrated tooth height, which causes the phase of the scattered light diffracting off different teeth to be randomly different. (ii) The reduction in energy flux of the scattered light due to the teeth reaching through six Fresnel zones, on average.]

(e) To aid you in answering part d, show that the propagator (point-spread function) for light that begins at the center of one mirror, travels to the edge of a baffle, and then propagates to the center of the distant mirror is

$$P \propto \int_o^{2\pi} \exp\left(\frac{ikR^2(\phi)}{2\ell_{\rm red}}\right) d\phi, \quad \text{where} \quad \frac{1}{\ell_{\rm red}} = \frac{1}{\ell} + \frac{1}{L - \ell}. \tag{8.44}$$

Here $R(\phi)$ is the radial distance of the baffle edge from the beam tube axis at azimuthal angle ϕ around the beam tube, and at distance ℓ down the tube from the scattering mirror. Note that $\ell_{\rm red}$ is the "reduced baffle distance" analogous to the reduced mass in a binary system. One can show that the time-varying part of the scattered-light amplitude (i.e., the part whose time dependence is produced by baffle vibrations) is proportional to this propagator. Explain why this is plausible. Then explain how the baffle serrations, embodied in the ϕ dependence of $R(\phi)$, produce the reduction of scattered-light amplitude in the manner described in part c.

8.6 Diffraction at a Caustic

In Sec. 7.5, we described how caustics can be formed in general in the geometric-optics limit (e.g., on the bottom of a swimming pool when the water's surface is randomly rippled, or behind a gravitational lens). We chose as an example an imperfect lens illuminated by a point source (Fig. 7.13) and showed how a pair of images would merge as the transverse distance x of the observer from the caustic decreases to zero. That merging was controlled by maxima and minima of the scaled time delay $\tilde{t}$ or equivalently, the phase φ of the light that originates at transverse location a just before the lens and arrives at transverse location x in the observation plane. We argued that for a fold caustic, this phase, when expressed as a Taylor series in a, has the standard form $\varphi(\tilde{a}, \tilde{x}) = \tilde{a}^3/3 - \tilde{x}\tilde{a}$, where $\tilde{a}$ and $\tilde{x}$ are rescaled a and x [Eq. (7.72)]. Using this φ, we showed that the magnification $\mathcal{M}$ of the images diverged $\propto \tilde{x}^{-1/2}$ as the caustic was approached ($\tilde{x} \to 0$), and then $\mathcal{M}$ crashed to zero just past the caustic (the two images disappeared). This singular behavior raised the question of what happens when we take into account the finite wavelength of the wave while still assuming the source has negligible size.

geometric optics divergence at a fold caustic

We are now in a position to answer this question. We simply use the Helmholtz-Kirchhoff formula (8.6) to write the expression for the amplitude measured at position $\tilde{x}$ in the form (Ex. 8.17)

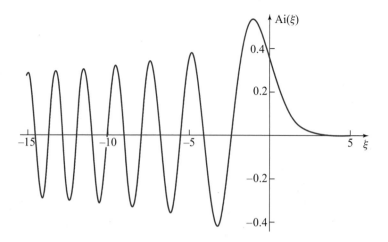

FIGURE 8.19 The Airy function $Ai(\xi)$ describing diffraction at a fold caustic. The argument is $\xi = -\tilde{x}$, where $\tilde{x}$ is rescaled distance from the caustic.

$$\psi(\tilde{x}) \propto \int d\tilde{a}\, e^{i\varphi(\tilde{a}, \tilde{x})} = \int d\tilde{a}(\cos\varphi + i\sin\varphi), \quad \text{where} \quad \varphi = \frac{\tilde{a}^3}{3} - \tilde{x}\tilde{a}.$$

(8.45)

The phase φ varies rapidly with location $\tilde{a}$ in the lens at large $|\tilde{a}|$, so we can treat the limits of integration as $\pm\infty$. Because $\varphi(\tilde{a}, \tilde{x})$ is odd in $\tilde{a}$, the sine term integrates to zero, and the cosine integral turns out to be the Airy function

diffraction at a caustic: Airy function

$$\psi \propto \int_{-\infty}^{\infty} d\tilde{a}\, \cos(\tilde{a}^3/3 - \tilde{x}\tilde{a}) = 2\pi\, Ai(-\tilde{x}).$$

(8.46)

The Airy function $Ai(\xi)$ is displayed in Fig. 8.19.

The asymptotic behavior of $Ai(\xi)$ is

$$Ai(\xi) \sim \pi^{-1/2}|\xi|^{-1/4}\sin(2|\xi|^{3/2}/3 + \pi/4), \quad \text{for } \xi \to -\infty$$

$$\sim \frac{e^{-2\xi^{3/2}/3}}{2\pi^{1/2}\xi^{1/4}}, \quad \text{for } \xi \to \infty.$$

(8.47)

diffractive field structure at a fold caustic

We see that the amplitude ψ remains finite as the caustic is approached (as $\tilde{x} = -\xi \to 0$) instead of diverging as in the geometric-optics limit, and it decreases smoothly toward zero when the caustic is passed, instead of crashing instantaneously to zero. For $\tilde{x} > 0$ ($\xi = -\tilde{x} < 0$; left part of Fig. 8.19), where an observer sees two geometric-optics images, the envelope of ψ diminishes $\propto \tilde{x}^{-1/4}$, so the energy flux $|\psi|^2$ decreases $\propto \tilde{x}^{-1/2}$ just as in the geometric-optics limit. What is actually seen is a series of bands of alternating dark and light with their spacing calculated by using $\Delta(2\tilde{x}^{3/2}/3) = \pi$ or $\Delta\tilde{x} \propto \tilde{x}^{-1/2}$. At sufficient distance from the caustic, it is not possible to resolve these bands, and a uniform illumination of average flux is observed, so we recover the geometric-optics limit.

These near-caustic scalings and others in Ex. 8.16, like the geometric-optics scalings (Sec. 7.5), are a universal property of this type of caustic (the simplest caustic of all, the "fold").

There is a helpful analogy, familiar from quantum mechanics. Consider a particle in a harmonic potential well in a highly excited state. Its wave function is given in the usual way using Hermite polynomials of large order. Close to the classical turning point, these functions change from being oscillatory to having exponential decay, just like the Airy function (and if we were to expand about the turning point, we would recover Airy functions). Of course, what is happening is that the probability density of finding the particle close to its turning point diverges classically, because it is moving vanishingly slowly at the turning point; the oscillations stem from interference between waves associated with the particle moving in opposite directions at the turning point.

analogy with quantum wave function in harmonic oscillator potential

For light near a caustic, if we consider the transverse component of the photon motion, then we have essentially the same problem. The field's oscillations stem from interference of the waves associated with the motions of the photons in two geometric-optics beams coming from slightly different directions and thus having slightly different transverse photon speeds.

This is our first illustration of the formation of large-contrast interference fringes when only a few beams are combined. We shall meet other examples of such interference in the next chapter.

EXERCISES

Exercise 8.16 *Problem: Wavelength Scaling at a Fold Caustic*
For the fold caustic discussed in the text, assume that the phase change introduced by the imperfect lens is nondispersive, so that the $\varphi(\tilde{a}, \tilde{x})$ in Eq. (8.45) satisfies $\varphi \propto \lambda^{-1}$. Show that the peak magnification of the interference fringes at the caustic scales with wavelength, $\propto \lambda^{-4/3}$. Also show that the spacing Δx of the fringes near a fixed observing position x is $\propto \lambda^{2/3}$. Discuss qualitatively how the fringes will be affected if the source has a finite size.

Exercise 8.17 *Problem: Diffraction at Generic Caustics*
In Sec. 7.5, we explored five elementary (generic) caustics that can occur in geometric optics. Each is described by its phase $\varphi(\tilde{a}, \tilde{b}; \tilde{x}, \tilde{y}, \tilde{z})$ for light arriving at an observation point with Cartesian coordinates $\{\tilde{x}, \tilde{y}, \tilde{z}\}$ along paths labeled by $(\tilde{a}, \tilde{b})$.

(a) Suppose the (monochromatic) wave field $\psi(\tilde{x}, \tilde{y}, \tilde{z})$ that exhibits one of these caustics is produced by plane-wave light that impinges orthogonally on a phase-shifting surface on which are laid out Cartesian coordinates $(\tilde{a}, \tilde{b})$. Using the Helmholtz-Kirchhoff diffraction integral (8.6), show that the field near a caustic is given by

$$\psi(\tilde{x}, \tilde{y}, \tilde{z}) \propto \int \int d\tilde{a} \, d\tilde{b} \, e^{i\varphi(\tilde{a}, \tilde{b}; \tilde{x}, \tilde{y}, \tilde{z})}. \qquad (8.48)$$

(b) In the text we evaluated this near-caustic diffraction integral for a fold caustic, obtaining the Airy function. For the higher-order elementary caustics, the integral cannot be evaluated analytically in terms of standard functions. To get insight into the influence of finite wavelength, evaluate the integral numerically for the case of a cusp caustic, $\varphi = \frac{1}{4}\tilde{a}^4 - \frac{1}{2}\tilde{z}\tilde{a}^2 - \tilde{x}\tilde{a}$, and plot the real and imaginary parts of ψ ($\Re\psi$ and $\Im\psi$). Before doing so, though, guess what these parts will look like. As foundations for this guess, (i) pay attention to the shape $\tilde{x} = \pm 2(\tilde{z}/3)^{3/2}$ of the caustic in the geometric-optics approximation, (ii) notice that away from the cusp point, each branch of the caustic is a fold, whose ψ is the Airy function (Fig. 8.19), and (iii) note that the oscillating ψ associated with each branch interferes with that associated with the other branch. The numerical computation may take awhile, so make a wise decision from the outset as to the range of $\tilde{z}$ and $\tilde{x}$ to include in your computations and plots. If your computer is really slow, you may want to prove that the integral is symmetric in $\tilde{x}$ and so restrict yourself to positive $\tilde{x}$, and argue that the qualitative behaviors of $\Re\psi$ and $\Im\psi$ must be the same, and so restrict yourself to $\Re\psi$.

Bibliographic Note

Hecht (2017) gives a pedagogically excellent treatment of diffraction at roughly the same level as this chapter, but in much more detail, with many illustrations and intuitive explanations. Other nice treatments at about our level will be found in standard optics textbooks, including Jenkins and White (1976), Brooker (2003), Sharma (2006), Bennett (2008), Ghatak (2010), and, from an earlier era, Longhurst (1973) and Welford (1988). The definitive treatment of diffraction at an advanced and thorough level is that of Born and Wolf (1999). For an excellent and thorough treatment of paraxial Fourier optics and spatial filtering, see Goodman (2005). The standard textbooks say little or nothing about diffraction at caustics, though they should; for this, we recommend the brief treatment by Berry and Upstill (1980) and the thorough treatment by Nye (1999).

Interference and Coherence

When two Undulations, from different Origins, coincide either perfectly or very nearly in Direction, their joint effect is a Combination of the Motions belonging to each.

THOMAS YOUNG (1802)

9.1 Overview

In the last chapter, we considered superpositions of waves that pass through a (typically large) aperture. The foundation for our analysis was the Helmholtz-Kirchhoff expression (8.4) for the field at a chosen point $\mathcal{P}$ as a sum of contributions from all points on a closed surface surrounding $\mathcal{P}$. The spatially varying field pattern resulting from this superposition of many different contributions is known as diffraction.

In this chapter, we continue our study of superposition, but for the special case where only two (or at most, several) discrete beams are being superposed. For this special case one uses the term *interference* rather than diffraction. Interference is important in a wide variety of practical instruments designed to measure or use the spatial and temporal structures of electromagnetic radiation. However, interference is not just of practical importance. Attempting to understand it forces us to devise ways of describing the radiation field that are independent of the field's origin and of the means by which it is probed. Such descriptions lead us naturally to the fundamental concept of coherence (Sec. 9.2).

The light from a distant, monochromatic point source is effectively a plane wave; we call it "perfectly coherent" radiation. In fact, there are two different types of coherence present: *lateral or spatial coherence* (coherence in the angular structure of the radiation field), and *temporal or longitudinal coherence* (coherence in the field's temporal structure, which clearly must imply something also about its frequency structure). We shall see in Sec. 9.2 that for both types of coherence there is a measurable quantity, called the *degree of coherence*, that is the Fourier transform of either the radiation's angular intensity distribution $I(\boldsymbol{\alpha})$ (energy flux per unit angle or solid angle, as a function of direction $\boldsymbol{\alpha}$) or its spectral energy flux $F_\omega(\omega)$ (energy flux per unit angular frequency ω, as a function of angular frequency).

Interspersed with our development of the theory of coherence are two historical devices with modern applications: (i) the stellar interferometer (Sec. 9.2.5), by which Michelson measured the diameters of Jupiter's moons and several bright stars using spatial coherence; and (ii) the Michelson interferometer and its practical implementation in a Fourier-transform spectrometer (Sec. 9.2.7), which use temporal coherence to measure electromagnetic spectra (e.g., the spectral energy flux of the cosmic microwave background radiation). After developing our full formalism for coherence, we go on in Sec. 9.3 to apply it to the operation of radio telescope arrays, which function by measuring the spatial coherence of the radiation field.

In Sec. 9.4, we turn to multiple-beam interferometry, in which incident radiation is split many times into several different paths and then recombined. A simple example is an etalon, made from two parallel, reflecting surfaces. A Fabry-Perot cavity interferometer, in which light is trapped between two highly reflecting mirrors (e.g., in a laser), is essentially a large-scale etalon. In Secs. 9.4.3 and 9.5, we discuss a number of applications of Fabry-Perot interferometers, including lasers, their stabilization, manipulations of laser light, the optical frequency comb, and laser interferometer gravitational-wave detectors.

Finally, in Sec. 9.6, we turn to the intensity interferometer. This has not proved especially powerful in application but does illustrate some quite subtle issues of physics and, in particular, highlights the relationship between the classical and quantum theories of light.

9.2 9.2 Coherence

9.2.1 9.2.1 Young's Slits

Young's slits

The most elementary example of interference is provided by Young's slits. Suppose two long, narrow, parallel slits are illuminated coherently by monochromatic light from a distant source that lies on the perpendicular bisector of the line joining the slits (the

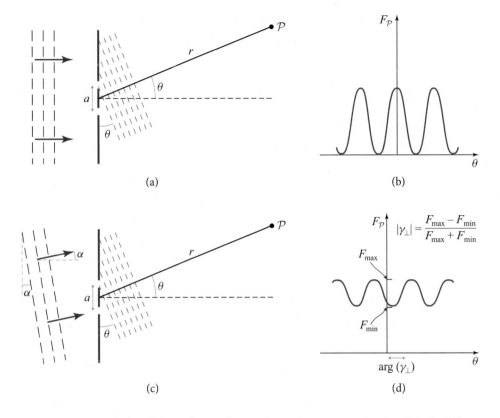

FIGURE 9.1 (a) Young's slits. (b) Interference fringes observed in a transverse plane [Eq. (9.1b)]. (c) The propagation direction of the incoming waves is rotated to make an angle α to the optic axis; as a result, the angular positions of the interference fringes in panel b are shifted by $\Delta\theta = \alpha$ [Eq. (9.3); not shown]. (d) Interference fringes observed from an extended source [Eq. (9.8)].

optic axis), so an incident wavefront reaches the slits simultaneously (Fig. 9.1a). This situation can be regarded as having only one lateral dimension because of translation invariance in the other. The waves from the slits (effectively, two 1-dimensional beams) fall onto a screen in the distant, Fraunhofer region, and there they interfere. The Fraunhofer interference pattern observed at a point $\mathcal{P}$, whose position is specified using the polar coordinates $\{r, \theta\}$ shown in Fig. 9.1, is proportional to the spatial Fourier transform of the transmission function [Eq. (8.11a)]. If the slits are very narrow, we can regard the transmission function as two delta functions, separated by the slit spacing a, and its Fourier transform will be

$$\psi(\theta) \propto e^{-ika\theta/2} + e^{ika\theta/2} \propto \cos\left(\frac{ka\theta}{2}\right), \tag{9.1a}$$

where $k = 2\pi/\lambda$ is the light's wave number, and a is the slit's separation. (That we can sum the wave fields from the two slits in this manner is a direct consequence of the linearity of the underlying wave equation.) The *energy flux* (energy per unit time crossing a unit area) at $\mathcal{P}$ (at angle θ to the optic axis) will be

$$\boxed{F(\theta) \propto |\psi|^2 \propto \cos^2(ka\theta/2);} \tag{9.1b}$$

interference fringes from Young's slits

cf. Fig. 9.1b. The alternating regions of dark and bright illumination in this flux distribution are known as *interference fringes*. Notice that the flux falls to zero between the bright fringes. This will be very nearly so even if (as is always the case in practice) the field hitting the slits has the form $e^{i[\omega_o t + \delta\varphi(t)]}$, where $\omega_o = c/k$ is the light's average angular frequency, and $\delta\varphi(t)$ is a phase [not to be confused with the light's full phase $\varphi = \omega_o t + \delta\varphi(t)$], which varies randomly on a timescale extremely long compared to $1/\omega_o$.[1] Notice also that there are many fringes, symmetrically disposed with respect to the optic axis. [If we were to take account of the finite width $w \ll a$ of the two slits, then we would find, by contrast with Eq. (9.1b), that the actual number of fringes is finite, in fact of order

interference by division of the wavefront

a/w; cf. Fig. 8.6 and associated discussion.] This type of interferometry is sometimes known as *interference by division of the wavefront*.

Young's slits in quantum mechanics

Of course, this Young's slits experiment is familiar from quantum mechanics, where it is often used as a striking example of the nonparticulate behavior of electrons (e.g., Feynman, Leighton, and Sands, 2013, Vol. III, Chap. 1). Just as for electrons, so also for photons it is possible to produce interference fringes even if only one photon is in the apparatus at any time, as was demonstrated in a famous experiment performed by G. I. Taylor in 1909. However, our concerns in this chapter are with the classical limit, where many photons are present simultaneously and their fields can be described by Maxwell's equations. In the next subsection we depart from the usual quantum mechanical treatment by asking what happens to the fringes when the source of radiation is spatially extended.

EXERCISES

Exercise 9.1 *Problem: Single-Mirror Interference*
X-rays with wavelength 8.33 Å (0.833 nm) coming from a point source can be reflected at shallow angles of incidence from a plane mirror. The direct ray from a point source

---

1. More precisely, if $\delta\varphi(t)$ wanders by $\sim\pi$ on a timescale $\tau_c \gg 2\pi/\omega_o$ (the waves' *coherence time*), then the waves are contained in a bandwidth $\Delta\omega_o \sim 2\pi/\tau_c \ll \omega_o$ centered on ω_o, k is in a band $\Delta k \sim k\Delta\omega/\omega_o$, and the resulting superposition of precisely monochromatic waves has fringe minima with fluxes $F_{\min}$ that are smaller than the maxima by $F_{\min}/F_{\max} \sim (\pi\Delta\omega/\omega_o)^2 \ll 1$. [One can see this in order of magnitude by superposing the flux (9.1b) with wave number k and the same flux with wave number $k + \Delta k$.] Throughout this section, until Eq. (9.15), we presume that the waves have such a small bandwidth (such a long coherence time) that this $F_{\min}/F_{\max}$ is completely negligible; for example, $1 - F_{\min}/F_{\max}$ is far closer to unity than any fringe visibility V [Eq. (9.8)] that is of interest to us. This can be achieved in practice by either controlling the waves' source or by band-pass filtering the measured signals just before detecting them.

to a detector 3 m away interferes with the reflected ray to produce fringes with spacing 25 μm. Calculate the distance of the X-ray source from the mirror plane.

9.2.2 Interference with an Extended Source: Van Cittert-Zernike Theorem

We approach the topic of extended sources in steps. Our first step was taken in the last subsection, where we dealt with an idealized, single, incident plane wave, such as might be produced by an ideal, distant laser. We have called this type of radiation "perfectly coherent," which we have implicitly taken to mean that the field oscillates with a fixed angular frequency ω_o and a randomly but very slowly varying phase $\delta\varphi(t)$ (see footnote 1), and thus, for all practical purposes, there is a time-independent phase difference between any two points in the region under consideration.

perfect coherence

As our second step, we keep the incoming waves perfectly coherent and perfectly planar, but change their incoming direction in Fig. 9.1 so it makes a small angle α to the optic axis (and correspondingly, its wavefronts make an angle α to the plane of the slits) as shown in Fig. 9.1c. Then the distribution of energy flux in the Fraunhofer diffraction pattern on the screen will be modified to

$$F(\theta) \propto |e^{-ika(\theta-\alpha)/2} + e^{+ika(\theta-\alpha)/2}|^2 \propto \cos^2\left(\frac{ka(\theta-\alpha)}{2}\right)$$

interference fringes for perfectly coherent waves from angle α

$$\propto \{1 + \cos[ka(\theta-\alpha)]\}. \tag{9.2}$$

Notice that, as the direction α of the incoming waves is varied, the locations of the bright and dark fringes change by $\Delta\theta = \alpha$, but the fringes remain fully sharp (their minima remain essentially zero; cf. footnote 1). Thus, the positions of the fringes carry information about the direction to the source.

Now, in our third and final step, we deal with an extended source (i.e., one whose radiation comes from a finite range of angles α), with (for simplicity) $|\alpha| \ll 1$. We assume that the source is monochromatic (and in practice we can make it very nearly monochromatic by band-pass filtering the waves just before detection). However, in keeping with how all realistic monochromatic sources (including band-pass filtered sources) behave, we give it a randomly fluctuating phase $\delta\varphi(t)$ [and amplitude $A(t)$], and require that the timescale on which the phase and amplitude wander (the waves' coherence time) be long compared to the waves' period $2\pi/\omega_o$; cf. footnote 1.

We shall also assume that the sources of the light propagating in different directions are independent and uncorrelated. Typically, they are separate electrons, ions, atoms, or molecules. To make this precise, we write the field in the form[2]

$$\Psi(x, z, t) = e^{i(kz-\omega_o t)} \int \psi(\alpha, t) e^{ik\alpha x} d\alpha, \tag{9.3}$$

wave field from extended source

---

2. As in Chap. 8, we denote the full field by Ψ and reserve ψ to denote the portion of the field from which a monochromatic part $e^{-i\omega_o t}$ or $e^{i(kz-\omega_o t)}$ has been factored out.

where $\psi(\alpha, t) = A e^{-i\delta\varphi}$ is the slowly wandering complex amplitude of the waves from direction α. When we consider the total flux arriving at a given point (x, z) from two different directions α_1 and α_2 and average it over times long compared to the waves' coherence time, then we lose all interference between the two contributions:

incoherent superposition of radiation

$$\overline{|\psi(\alpha_1, t) + \psi(\alpha_2, t)|^2} = \overline{|\psi(\alpha_1, t)|^2} + \overline{|\psi(\alpha_2, t)|^2}. \tag{9.4}$$

Such radiation is said to be incoherent in the incoming angle α, and we say that the contributions from different directions *superpose incoherently.* This is just a fancy way of saying that their intensities (averaged over time) add linearly.

The angularly incoherent light from our extended source is sent through two Young's slits and produces fringes on a screen in the distant Fraunhofer region. We assume that the coherence time for the light from each source point is very long compared to the difference in light travel time to the screen via the two different slits. Then the light from each source point in the extended source forms the sharp interference fringes described by Eq. (9.2). However, because contributions from different source directions add incoherently, the flux distribution on the screen is a linear sum of the fluxes from source points:

$$F(\theta) \propto \int d\alpha\, I(\alpha)\{1 + \cos[ka(\theta - \alpha)]\}. \tag{9.5}$$

Here $I(\alpha)d\alpha \propto \overline{|\psi(\alpha, t)|^2}d\alpha$ is the flux incident on the plane of the slits from the infinitesimal range $d\alpha$ of directions, so $I(\alpha)$ is the radiation's *intensity* (its energy per unit time falling on a unit area and coming from a unit angle). The remainder of the integrand, $1 + \cos[ka(\theta - \alpha)]$, is the Fraunhofer diffraction pattern [Eq. (9.2)] for coherent radiation from direction α.

We presume that the range of angles present in the waves, $\Delta\alpha$, is large compared to their fractional bandwidth $\Delta\alpha \gg \Delta\omega/\omega_o$; so, whereas the finite but tiny bandwidth produced negligible smearing out of the interference fringes (see footnote 1 in this chapter), the finite but small range of directions may produce significant smearing [i.e., the minima of $F(\theta)$ might not be very sharp]. We quantify the fringes' non-sharpness and their locations by writing the slit-produced flux distribution (9.5) in the form

interference fringes from extended source

$$\boxed{F(\theta) = F_S[1 + \Re\{\gamma_\perp(ka)e^{-ika\theta}\}],} \tag{9.6a}$$

where

$$\boxed{F_S \equiv \int d\alpha\, I(\alpha)} \tag{9.6b}$$

(subscript S for "source") is the total flux arriving at the slits from the source, and

degree of lateral coherence

$$\boxed{\gamma_\perp(ka) \equiv \frac{\int d\alpha\, I(\alpha)e^{ika\alpha}}{F_S}} \tag{9.7a}$$

is defined as the radiation's *degree of spatial (or lateral) coherence*.[3] The phase of $\gamma_\perp$ determines the angular locations of the fringes; its modulus determines their depth (the amount of their smearing due to the source's finite angular size).

The nonzero value of $\gamma_\perp(ka)$ reflects the existence of some degree of relative coherence between the waves arriving at the two slits, whose separation is a. The radiation can have this finite spatial coherence, despite its complete lack of angular coherence, because each angle contributes coherently to the field at the two slits. The lack of coherence for different angles reduces the net spatial coherence (smears the fringes), but it does not drive the coherence all the way to zero (does not completely destroy the fringes).

Equation (9.7a) states that the degree of lateral coherence of the radiation from an extended, angularly incoherent source is the Fourier transform of the source's angular intensity pattern. Correspondingly, if one knows the degree of lateral coherence as a function of the (dimensionless) distance ka, from it one can reconstruct the source's angular intensity pattern by Fourier inversion:

$$I(\alpha) = F_S \int \frac{d(ka)}{2\pi} \gamma_\perp(ka) e^{-ika\alpha}. \tag{9.7b}$$

The two Fourier relations (9.7a) and (9.7b) make up the *van Cittert-Zernike theorem*. In Ex. 9.8, we shall see that this theorem is a complex-variable version of Chap. 6's *Wiener-Khintchine theorem* for random processes.

van Cittert-Zernike theorem for lateral coherence

Because of its Fourier-transform relationship to the source's angular intensity pattern $I(\alpha)$, the degree of spatial coherence $\gamma_\perp(ka)$ is of great practical importance. For a given choice of ka (a given distance between the slits), $\gamma_\perp$ is a complex number that one can read off the interference fringes of Eq. (9.6a) and Fig. 9.1d as follows. Its modulus is

$$|\gamma_\perp| \equiv V = \frac{F_{\max} - F_{\min}}{F_{\max} + F_{\min}}, \tag{9.8}$$

where $F_{\max}$ and $F_{\min}$ are the maximum and minimum values of the flux F on the screen; and its phase $\arg(\gamma_\perp)$ is ka times the displacement $\Delta\theta$ of the centers of the bright fringes from the optic axis. The modulus (9.8) is called the *fringe visibility*, or simply the *visibility*, because it measures the fractional contrast in the fringes [Eq. (9.8)], and this name is the reason for the symbol V. Analogously, the complex quantity $\gamma_\perp$ (or a close relative) is sometimes known as the *complex fringe visibility*. Notice that V can lie anywhere in the range from zero (no contrast; fringes completely undetectable) to unity (monochromatic plane wave; contrast as large as possible).

fringe visibility, or visibility

complex fringe visibility

---

3. In Sec. 9.2.6, we introduce the degree of temporal coherence (also known as longitudinal coherence) to describe the correlation along the direction of propagation. The correlation in a general, lateral and longitudinal, direction is called the degree of coherence (Sec. 9.2.8).

When the phase $\arg(\gamma_\perp)$ of the complex visibility (degree of coherence) is zero, there is a bright fringe precisely on the optic axis. This will be the case for a source that is symmetric about the optic axis, for example. If the symmetry point of such a source is gradually moved off the optic axis by an angle $\delta\alpha$, the fringe pattern will shift correspondingly by $\delta\theta = \delta\alpha$, which will show up as a corresponding shift in the argument of the fringe visibility, $\arg(\gamma_\perp) = ka\delta\alpha$.

The above analysis shows that Young's slits, even when used virtually, are nicely suited to measuring both the modulus and the phase of the complex fringe visibility (the degree of spatial coherence) of the radiation from an extended source.

9.2.3

9.2.3 More General Formulation of Spatial Coherence; Lateral Coherence Length

It is not necessary to project the light onto a screen to determine the contrast and angular positions of the fringes. For example, if we had measured the field at the locations of the two slits, we could have combined the signals electronically and cross correlated them numerically to determine what the fringe pattern would be with slits. All we are doing with the Young's slits is sampling the wave field at two different points, which we now label 1 and 2. Observing the fringes corresponds to adding a phase φ ($= ka\theta$) to the field at one of the points and then adding the fields and measuring the flux $\propto |\psi_1 + \psi_2 e^{i\varphi}|^2$ averaged over many periods. Now, since the source is far away, the rms value of the wave field will be the same at the two slits: $\overline{|\psi_1|^2} = \overline{|\psi_2|^2} \equiv \overline{|\psi|^2}$. We can therefore express this time-averaged flux in the symmetric-looking form

$$F(\varphi) \propto \overline{(\psi_1 + \psi_2 e^{i\varphi})(\psi_1^* + \psi_2^* e^{-i\varphi})}$$

$$\propto 1 + \Re\left(\frac{\overline{\psi_1 \psi_2^*}}{\overline{|\psi|^2}} e^{-i\varphi}\right). \tag{9.9}$$

degree of spatial coherence

Here a bar denotes an average over times long compared to the coherence times for ψ_1 and ψ_2. Comparing with Eq. (9.6a) and using $\varphi = ka\theta$, we identify

$$\gamma_{\perp 12} = \frac{\overline{\psi_1 \psi_2^*}}{\overline{|\psi|^2}} \tag{9.10}$$

as the *degree of spatial coherence* in the radiation field between the two points 1 and 2. Equation (9.10) is the general definition of degree of spatial coherence. Equation (9.6a) is the special case for points separated by a lateral distance a.

If the radiation field is strongly correlated between the two points, we describe it as having strong spatial or lateral coherence. Correspondingly, we shall define a field's

spatial or lateral coherence length

lateral coherence length $l_\perp$ as the linear size of a region over which the field is strongly correlated (has $V = |\gamma_\perp| \sim 1$). If the angle subtended by the source is $\sim\delta\alpha$, then by virtue of the van Cittert-Zernike theorem [Eqs. (9.7)] and the usual reciprocal relation for Fourier transforms, the radiation field's lateral coherence length will be

Chapter 9. Interference and Coherence

$$l_\perp \sim \frac{2\pi}{k\,\delta\alpha} = \frac{\lambda}{\delta\alpha}. \tag{9.11}$$

This relation has a simple physical interpretation. Consider two beams of radiation coming from opposite sides of the brightest portion of the source. These beams are separated by the incoming angle $\delta\alpha$. As one moves laterally in the plane of the Young's slits, one sees a varying relative phase delay between these two beams. The coherence length $l_\perp$ is the distance over which the variations in that relative phase delay are of order 2π: $k\,\delta\alpha\, l_\perp \sim 2\pi$.

9.2.4 Generalization to 2 Dimensions

We have so far just considered a 1-dimensional intensity distribution $I(\alpha)$ observed through the familiar Young's slits. However, most sources will be 2-dimensional, so to investigate the full radiation pattern, we should allow the waves to come from 2-dimensional angular directions $\boldsymbol{\alpha}$:

$$\Psi = e^{i(kz-\omega_0 t)} \int \psi(\boldsymbol{\alpha}, t) e^{ik\boldsymbol{\alpha}\cdot\mathbf{x}} d^2\alpha \equiv e^{i(kz-\omega_0 t)} \psi(\mathbf{x}, t) \tag{9.12a}$$

[where $\psi(\boldsymbol{\alpha}, t)$ is slowly varying in time], and we should use several pairs of slits aligned along different directions. Stated more generally, we should sample the wave field (9.12a) at a variety of points separated by a variety of 2-dimensional vectors $\mathbf{a}$ transverse to the direction of wave propagation. The complex visibility (degree of spatial coherence) will then be a function of $k\mathbf{a}$,

$$\gamma_\perp(k\mathbf{a}) = \frac{\overline{\psi(\mathbf{x}, t)\psi^*(\mathbf{x} + \mathbf{a}, t)}}{\overline{|\psi|^2}}, \tag{9.12b}$$

complex visibility, or degree of spatial coherence

and the van Cittert-Zernike theorem (9.7) will take the 2-dimensional form

$$\gamma_\perp(k\mathbf{a}) = \frac{\int d\Omega_\alpha I(\boldsymbol{\alpha}) e^{ik\mathbf{a}\cdot\boldsymbol{\alpha}}}{F_S}, \tag{9.13a}$$

2-dimensional van Cittert-Zernike theorem

$$I(\boldsymbol{\alpha}) = F_S \int \frac{d^2(ka)}{(2\pi)^2} \gamma_\perp(k\mathbf{a}) e^{-ik\mathbf{a}\cdot\boldsymbol{\alpha}}. \tag{9.13b}$$

Here $I(\boldsymbol{\alpha}) \propto \overline{|\psi(\boldsymbol{\alpha}, t)|^2}$ is the source's *intensity* (energy per unit time crossing a unit area from a unit solid angle $d\Omega_\alpha$); $F_S = \int d\Omega_\alpha I(\boldsymbol{\alpha})$ is the source's total energy flux; and $d^2(ka) = k^2 d\Sigma_a$ is a (dimensionless) surface area element in the lateral plane.

EXERCISES

Exercise 9.2 *Problem: Lateral Coherence of Solar Radiation*
How closely separated must a pair of Young's slits be to see strong fringes from the Sun (angular diameter $\sim 0.5°$) at visual wavelengths? Suppose that this condition is

just satisfied, and the slits are 10 μm in width. Roughly how many fringes would you expect to see?

Exercise 9.3 *Problem: Degree of Coherence for a Source with Gaussian Intensity Distribution*
A circularly symmetric light source has an intensity distribution $I(\alpha) = I_0 \exp[-\alpha^2/(2\alpha_0^2)]$, where α is the angular radius measured from the optic axis. Compute the degree of spatial coherence. What is the lateral coherence length? What happens to the degree of spatial coherence and the interference fringe pattern if the source is displaced from the optic axis?

9.2.5 Michelson Stellar Interferometer; Astronomical Seeing

The classic implementation of Young's slits for measuring spatial coherence is Michelson's stellar interferometer, which Albert Michelson and Francis Pease (1921) used for measuring the angular diameters of Betelgeuse and several other bright stars in 1920.[4] The starlight was sampled at two small mirrors separated by a variable distance $a \leq 6$ m and was then reflected into the 100-inch (2.5-m) telescope on Mount Wilson, California, to form interference fringes (Fig. 9.2). (As we have emphasized, the way in which the fringes are formed is unimportant; all that matters is the two locations where the light is sampled, i.e., the first two mirrors in Fig. 9.2.) As Michelson and Pease increased the separation a between the mirrors, the fringe visibility V decreased. Michelson and Pease modeled Betelgeuse (rather badly, in fact) as a circular disk of uniform brightness, $I(\alpha) = 1$ for $|\alpha| < \alpha_r$ and 0 for $|\alpha| > \alpha_r$, so its visibility was given, according to Eq. (9.13a), as

$$V = \gamma_\perp = 2\,\text{jinc}(ka\alpha_r) \tag{9.14}$$

where α_r is the star's true angular radius, and $\text{jinc}(\xi) = J_1(\xi)/\xi$. They identified the separation $a \simeq 3$ m, where the fringes disappeared, with the first zero of the function $\text{jinc}(ka\alpha_r)$. From this and the mean wavelength $\lambda = 575$ nm of the starlight, they inferred that the angular radius of Betelgeuse is $\alpha_r \sim 0.02$ arcsec, which at Betelgeuse's then-estimated distance of 60 pc (180 light-years) corresponds to a physical radius $\sim$300 times larger than that of the Sun. The modern parallax-measured distance is 200 pc, so Betelgeuse's physical radius is actually $\sim$1,000 times larger than the Sun.

This technique only works for big, bright stars and is very difficult to use, because turbulence in Earth's atmosphere causes the fringes to keep moving around; see Box 9.2 and Ex. 9.4 for details.

---

4. Similar principles are relevant to imaging by Earth-observing satellites where one looks down, not up. In this case, there is the important distinction that most of the diffraction happens relatively close to the source, not the detector.

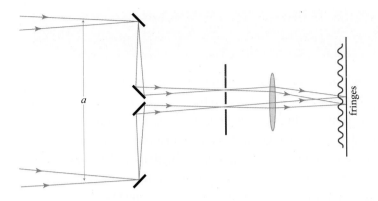

FIGURE 9.2 Schematic illustration of a Michelson stellar interferometer.

Exercise 9.4 *Example and Derivation: Time-Averaged Visibility and Image for a Distant Star Seen through Earth's Turbulent Atmosphere* T2

Fill in the details of the analysis of time-averaged seeing in Box 9.2. More specifically, do the following. If you have difficulty, Roddier (1981) may be helpful.

(a) Give an order-of-magnitude derivation of Eq. (4a) in Box 9.2 for the mean-square phase fluctuation of light induced by propagation through a thin, turbulent layer of atmosphere. [Hint: Consider turbulent cells of size a, each of which produces some $\delta\varphi$, and argue that the contributions add up as a random walk.]

(b) Deduce the factor 2.91 in Eq. (4a) in Box 9.2 by evaluating

$$D_{\delta\varphi} = k^2 \left\langle \left\{ \int_z^{z+\delta h} [\delta n(\mathbf{x}+a, z, t) - \delta n(\mathbf{x}, z, t)]\, dz \right\}^2 \right\rangle.$$

(c) Derive Eq. (4b) in Box 9.2 for the time-averaged complex visibility after propagating through the thin layer. [Hint: Because $\zeta \equiv \delta\varphi(\mathbf{x}, t) - \delta\varphi(\mathbf{x}+\mathbf{a}, t)$ is the result of contributions from a huge number of independent turbulent cells, the central limit theorem (Sec. 6.3.2) suggests it is a Gaussian random variable though, in fact intermittency (Sec. 15.3) can make it rather non-Gaussian. Idealizing it as Gaussian, evaluate $\gamma_\perp = \langle e^{i\zeta}\rangle = \langle\int_{-\infty}^{\infty} p(\zeta)e^{i\zeta}\,d\zeta\rangle$ with $p(\zeta)$ the Gaussian distribution.]

(d) Use the point-spread function (8.28) for free propagation of the light field ψ to show that, under free propagation, the complex visibility $\gamma_\perp(\mathbf{a}, z, t) = \langle\psi(\mathbf{x}+\mathbf{a}, z, t)\psi^*(\mathbf{x}, z, t)\rangle$ (with averaging over $\mathbf{x}$ and t) is constant (i.e., independent of height z).

(e) By combining parts c and d, deduce Eqs. (5) in Box 9.2 for the mean-square phase fluctuations and spacetime-averaged visibility on the ground.

(f) Perform a numerical Fourier transform of $\bar{\gamma}_\perp(\mathbf{a})$ [Eq. (5b) in Box 9.2] to get the time-averaged intensity distribution $I(\alpha)$. Construct a log-log plot of it, and compare with panel b of the first figure in Box 9.2. What is r_o for the observational data shown in that figure?

BOX 9.2. ASTRONOMICAL SEEING, SPECKLE IMAGE PROCESSING, AND ADAPTIVE OPTICS T2

When light from a star passes through turbulent layers of Earth's atmosphere, the turbulently varying index of refraction $n(\mathbf{x}, t)$ diffracts the light in a random, time varying way. One result is "twinkling" (fluctuations in the flux observed by eye on the ground, with fluctuational frequencies $f_o \sim 100$ Hz). Another is astronomical seeing: the production of many images of the star (i.e., *speckles*) as seen through a large optical telescope (panel a of the box figure), with the image pattern fluctuating at ~ 100 Hz.

Here and in Ex. 9.4 we quantify astronomical seeing using the theory of 2-dimensional lateral coherence. We do this not because seeing is important (though it is), but rather because our analysis provides an excellent illustration of three fundamental concepts working together: (i) turbulence in fluids and its Kolmogorov spectrum (Chap. 15), (ii) random processes (Chap. 6), and (iii) coherence of light (this chapter).

We begin by deriving, for a star with arbitrarily small angular diameter, the time-averaged complex visibility $\gamma_\perp$ observed on the ground and the visibility's Fourier transform [the observed intensity distribution averaged

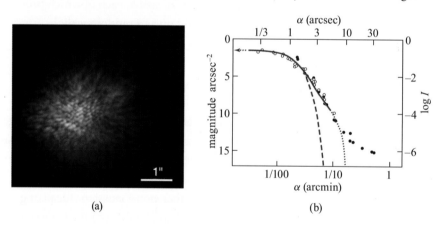

(a)   (b)

(a) Picture of a bright star with a dimmer companion star, as seen through the Russian 6-m telescope in the Caucasus Mountains, in an exposure shorter than 10 ms. Atmospheric turbulence creates a large number of images of each star (speckles) spread over a region with angular diameter of order 2 arcsec. (b) The theory discussed in the text and in Ex. 9.4 predicts the solid curve for the time-averaged intensity distribution for a single bright star. Notice the logarithmic axes. The dotted curve is an estimate of the influence of the small-scale cutoff of the turbulence, and the dashed curve is a Gaussian. The circles are observational data. Panel (a), Gerd Weigelt; panel (b), adapted from Roddier (1981).

(continued)

BOX 9.2. (continued)

over the speckles, $\bar{I}(\boldsymbol{\alpha})$]. Then we briefly discuss the temporally fluctuating speckle pattern and techniques for extracting information from it.

TIME-AVERAGED VISIBILITY AND ANGULAR DISTRIBUTION OF INTENSITY

When analyzing light propagation through a turbulent atmosphere, it is convenient to describe the turbulent fluctuations of the index of refraction by their spatial correlation function $C_{\mathfrak{n}}(\boldsymbol{\xi}) \equiv \langle \delta\mathfrak{n}(\mathbf{X}, t)\delta\mathfrak{n}(\mathbf{X} + \boldsymbol{\xi}, t) \rangle$ (discussed in Sec. 6.4.1); or, better yet, by $\mathfrak{n}$'s *mean-square fluctuation on the lengthscale* ξ,

$$D_{\mathfrak{n}}(\boldsymbol{\xi}) \equiv \langle [\delta\mathfrak{n}(\mathbf{X} + \boldsymbol{\xi}, t) - \delta\mathfrak{n}(\mathbf{X}, t)]^2 \rangle = 2[\sigma_{\mathfrak{n}}^2 - C_{\mathfrak{n}}(\boldsymbol{\xi})], \qquad (1)$$

which is called $\mathfrak{n}$'s *structure function*. Here $\delta\mathfrak{n}$ is the perturbation of the index of refraction, $\mathbf{X}$ is location in 3-dimensional space, t is time, $\langle \cdot \rangle$ denotes a spacetime average, and $\sigma_{\mathfrak{n}}^2 \equiv \langle \delta\mathfrak{n}^2 \rangle = C_{\mathfrak{n}}(0)$ is the variance of the fluctuations.

In Sec. 15.4.4, we show that, for strong and isotropic turbulence, $D_{\mathfrak{n}}$ has the functional form $D_{\mathfrak{n}} \propto \xi^{2/3}$ (where $\xi \equiv |\boldsymbol{\xi}|$), with a multiplicative coefficient $d_{\mathfrak{n}}^2$ that characterizes the strength of the perturbations:

$$D_{\mathfrak{n}}(\boldsymbol{\xi}) = d_{\mathfrak{n}}^2 \xi^{2/3} \qquad (2)$$

[Eq. (15.29)]. The 2/3 power is called the *Kolmogorov spectrum* for the turbulence.

When light from a very distant star (a point source), directly overhead for simplicity, reaches Earth's atmosphere, its phase fronts lie in horizontal planes, so the frequency ω component of the electric field is $\psi = e^{ikz}$, where z increases downward. (Here we have factored out the field's overall amplitude.) When propagating through a thin layer of turbulent atmosphere of thickness δh, the light acquires the phase fluctuation

$$\delta\varphi(\mathbf{x}, t) = k \int_z^{z+\delta h} \delta\mathfrak{n}(\mathbf{x}, z, t)dz. \qquad (3)$$

Here $\mathbf{x}$ is the transverse (i.e., horizontal) location, and Eq. (3) follows from $d\varphi = kdz$, with $k = (\mathfrak{n}/c)\omega$ and $\mathfrak{n} \simeq 1$.

In Ex. 9.4, we derive some spacetime-averaged consequences of the phase fluctuations (3):

1. When the light emerges from the thin, turbulent layer, it has acquired a mean-square phase fluctuation on transverse lengthscale a given by

$$D_{\delta\varphi}(\mathbf{a}) \equiv \langle [\delta\varphi(\mathbf{x} + \mathbf{a}, t) - \delta\varphi(\mathbf{x}, t)]^2 \rangle = 2.91 \, d_{\mathfrak{n}}^2 \, \delta h \, k^2 a^{5/3} \quad (4a)$$

(continued)

BOX 9.2. (continued)

[Eq. (2)], and a spacetime-averaged complex visibility given by

$$\bar{\gamma}_\perp(\mathbf{a}) = \langle \psi(\mathbf{x}, t)\psi^*(\mathbf{x} + \mathbf{a}, t)\rangle = \langle \exp\{i\,[\delta\varphi(\mathbf{x}, t) - \delta\varphi(\mathbf{x} + \mathbf{a}, t)]\}\rangle$$

$$= \exp\left[-\frac{1}{2}D_{\delta\varphi}(a)\right] = \exp\left[-1.455\,d_n^2\,\delta h k^2 a^{5/3}\right]. \quad (4b)$$

2. Free propagation (including free-propagator diffraction effects, which are important for long-distance propagation) preserves the spacetime-averaged complex visibility: $d\bar{\gamma}_\perp/dz = 0$.

3. Therefore, not surprisingly, when the turbulence is spread out vertically in some arbitrary manner, the net mean-square phase shift and time-averaged complex visibility observed on the ground are $D_\varphi(\mathbf{a}) = 2.91\left[\int d_n^2(z)dz\right]k^2a^{5/3}$, and $\bar{\gamma}_\perp(\mathbf{a}) = \exp\left[-\frac{1}{2}D_\varphi(a)\right]$.

It is conventional to introduce a transverse lengthscale

$$r_o \equiv \left[0.423k^2\int d_n^2(z)dz\right]^{-3/5}$$

called the *Fried parameter*, in terms of which D_φ and $\bar{\gamma}_\perp$ are

$$D_\varphi(\mathbf{a}) = 6.88(a/r_o)^{5/3}, \quad (5a)$$

$$\bar{\gamma}_\perp(\mathbf{a}) = \exp\left[-\frac{1}{2}D_\varphi(a)\right] = \exp\left[-3.44(a/r_o)^{5/3}\right]. \quad (5b)$$

This remarkably simple result provides opportunities to test the Kolmogorov power law. For light from a distant star, one can use a large telescope to measure $\bar{\gamma}_\perp(a)$ and then plot $\ln\bar{\gamma}_\perp$ as a function of a. Equation (5b) predicts a slope 5/3 for this plot, and observations confirm that prediction.

Notice that the Fried parameter r_o is the lengthscale on which the rms phase fluctuation $\varphi_{rms} = \sqrt{D_\varphi(a = r_o)}$ is $\sqrt{6.88} = 2.62$ radians: r_o is the transverse lengthscale beyond which the turbulence-induced phase fluctuations are large compared to unity. These large random phase fluctuations drive $\bar{\gamma}_\perp$ rapidly toward zero with increasing distance a [Eq. (5b)], that is, they cause the light field to become spatially decorrelated with itself for distances $a \gtrsim r_o$. Therefore, r_o is (approximately) the time-averaged light field's spatial correlation length on the ground. Moreover, since $\bar{\gamma}_\perp$ is preserved under free propagation from the turbulent region to the ground, r_o must be the transverse correlation length of the light as it exits the turbulent region that

(continued)

BOX 9.2. (continued)

produces the seeing. A correlated region with transverse size r_o is called an *isoplanatic patch*.

The observed time-averaged intensity $\bar{I}(\alpha)$ from the point-source star is the Fourier transform of the complex visibility (5b); see Eq. (9.13b). This transform cannot be performed analytically, but a numerical computation gives the solid curve in panel b of the figure above, which agrees remarkably well with observations out to $\sim 10^{-4}$ of the central intensity, where the Kolmogorov power law is expected to break down. Notice that the intensity distribution has a large-radius tail with far larger intensity than a Gaussian distribution (the dashed curve). This large-radius light is produced by large-angle diffraction, which is caused by very small-spatial-scale fluctuations (eddies) in the index of refraction.

Astronomers attribute to this time-averaged $I(\alpha)$ a full width at half maximum (FWHM) angular diameter $\alpha_{\text{Kol}}^{\text{FWHM}} = 0.98\lambda/r_o$ (Ex. 9.4). For blue light in very good seeing conditions, r_o is about 20 cm and $\alpha_{\text{Kol}}^{\text{FWHM}}$ is about 0.5 arcsec. Much more common is $r_o \sim 10$ cm and $\alpha_{\text{Kol}}^{\text{FWHM}} \sim 1$ arcsec.

SPECKLE PATTERN AND ITS INFORMATION

The speckle pattern seen on short timescales, $\lesssim 1/f_o \sim 0.01$ s, can be understood in terms of the turbulence's isoplanatic patches (see the drawing below). When the light field exits the turbulent region, at a height $h \lesssim 1$ km, the isoplanatic patches on its wavefronts, with transverse size r_o, are planar to within roughly a reduced wavelength $\lambda\!\!\!^- = 1/k = \lambda/(2\pi)$ (since the rms phase variation across a patch is just 2.62 radians). Each patch carries an image of the star, or whatever other object the astronomer is observing. The patch's light rays make an angle $\theta \lesssim \alpha_{\text{Kol}}^{\text{FWHM}} = 0.98\lambda/r_o$ to the vertical. The patch's Fresnel length from the ground is $r_F = \sqrt{\lambda\!\!\!^- h} \lesssim 2$ cm (since $\lambda\!\!\!^- \sim 0.5\,\mu$m and $h \lesssim 1$ km). This is significantly smaller than the patch size $r_o \sim 10$ to 20 cm; so there is little diffraction in the trip to ground. When these patches reach

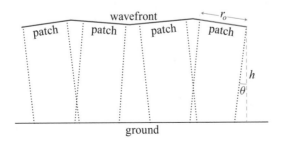

(continued)

BOX 9.2. (continued)

a large telescope (one with diameter $D \gg r_o$), each is focused to produce an image of the object centered on the angular position θ of its rays (dotted lines in the drawing). These images are the speckles seen in panel a of the first box figure.

The speckle pattern varies rapidly, because winds at high altitude sweep the isoplanatic patches through the star's light rays. For a wind speed $u \sim 20$ m s^{-1}, the frequency of the pattern's fluctuations is $f_o \sim u/r_o \sim 100$ Hz, in agreement with observations.

To study the speckles and extract information about the object's above-atmosphere intensity distribution $I_o(\boldsymbol{\alpha})$, one must observe them on timescales $\lesssim 1/f_o \sim 10$ ms. The first observations of this sort were the measurements of a few stellar diameters by Michelson and Pease, using the Michelson stellar interferometer (Sec. 9.2.7). The fringes they saw were produced by the speckles, and because the phase of each speckle's fringes was random, the many speckles contributed incoherently to produce a far smaller net fringe visibility V than in the absence of atmospheric turbulence. Moreover, because the speckle pattern varied at $f_o \sim 100$ Hz, the net visibility and its phase also varied at $f_o \sim 100$ Hz. Fortunately, the human eye can discern things that vary this fast, so Michelson and Pease were able to see the fringes.

In the modern era of CCDs, fast electronics, and powerful computers, a variety of more sophisticated techniques have been devised and implemented for observing these speckles and extracting their information. Two common techniques are speckle-image processing and adaptive optics.

In *speckle image processing,* which is really only usable for bright sources, one makes optical measurements of the speckle pattern (sometimes with multi-pinhole masks) on timescales $\lesssim 0.01$ s for which the speckles are unchanging. One then uses optical or computational techniques to construct fourth-order or sixth-order correlations of the light field, for example, $\int \gamma_\perp(\mathbf{a} - \mathbf{a}')\gamma_\perp^*(\mathbf{a}')d^2a'$ (which is fourth-order in the field), from which a good approximation to the source's above-atmosphere intensity distribution $I_o(\boldsymbol{\alpha})$ can be computed.

In *adaptive optics,* one focuses not on the speckles themselves, but on the turbulence-induced distortions of the wavefronts arriving at the large telescope's mirror. The simplest implementation uses *natural guide stars* to act as point sources. Unfortunately the incidence of sufficiently bright stars only allows a small fraction of the sky to be examined using this technique.

(continued)

BOX 9.2. (continued)

Therefore, *laser guide stars* are created by shining collimated laser light close to the direction of the astronomical sources under study. The laser light generates the guide star via either Rayleigh scattering by molecules at modest altitude ($\sim$20 km) or resonant scattering from a layer of sodium atoms at high altitude ($\sim$90 km).

The guide star must be within an angular distance $\lesssim r_o/h \sim 3$ arcsec of the astronomical object one is observing (where $h \sim 10$ km is the height of the highest turbulent layers that contribute significantly to the seeing). This $\lesssim 3$ arcsec separation guarantees that light rays arriving at the same spot on the telescope mirror from the object and from the artificial star will have traversed the same isoplanatic patches of turbulent atmosphere and thus have experienced the same phase delay and acquired the same wavefront distortions. One measures the wavefront distortions of the artificial star's light and dynamically reshapes the telescope mirror so as to compensate for them. This removes the distortions not only from the artificial star's wavefronts but also from the astronomical object's wavefronts. Thereby one converts the speckle pattern into the object's true intensity distribution $I_o(\boldsymbol{\alpha})$.

Two recent successes of adaptive optics are to observe the stars orbiting the massive black hole in our galactic center and thereby measure its mass (Ghez et al., 2008; Genzel, Eisenhauer, and Gillessen, 2010); and to image exoplanets directly by masking out the light—typically millions of times brighter than the planet—from the stars that they orbit (Macintosh et al., 2014).

The techniques described here also find application to the propagation of radio waves through the turbulent interplanetary and interstellar media. The refractive index is due to the presence of free electrons in a plasma (see Sec. 21.4.1). Interestingly, this turbulence is commonly characterized by a Kolmogorov spectrum, though the presence of a magnetic field makes it anisotropic.

These techniques are also starting to find application in ophthalmology and industry.

(g) Reexpress the turbulence-broadened image's FWHM as $\alpha_{\mathrm{Kol}}^{\mathrm{FWHM}} = 0.98\lambda/r_o$. Show that the Airy intensity distribution for light for a circular aperture of diameter D [Eq. (8.18)] has FWHM $\alpha_{\mathrm{Airy}}^{\mathrm{FWHM}} = 1.03\lambda/D$.

(h) The fact that the coefficients in these two expressions for α^{FWHM} are both close to unity implies that when the diameter $D \lesssim r_o$, the seeing is determined by the telescope. Conversely, when $D \gtrsim r_o$ (which is true for essentially all ground-based research optical telescopes), it is the atmosphere that determines the image quality. Large telescopes act only as "light buckets," unless some additional correctives are applied, such as speckle image processing or adaptive optics (Box 9.2). A common measure of the performance of a telescope with or without this correction is the *Strehl ratio*, S, which is the ratio of the peak intensity in the actual image of a point source on the telescope's optic axis to the peak intensity in the Airy disk for the telescope's aperture. Show that without correction, $S = 1.00(r_o/D)^2$. Modern adaptive optics systems on large telescopes can achieve $S \sim 0.5$.

9.2.6

9.2.6 Temporal Coherence

In addition to the degree of spatial (or lateral) coherence, which measures the correlation of the field transverse to the direction of wave propagation, we can also measure the *degree of temporal coherence*, also called the *degree of longitudinal coherence*. This describes the correlation at a given time at two points separated by a distance s along the direction of propagation. Equivalently, it measures the field sampled at a fixed position at two times differing by $\tau = s/c$. When (as in our discussion of spatial coherence) the waves are nearly monochromatic so the field arriving at the fixed position has the form $\Psi = \psi(t)e^{-i\omega_o t}$, then the degree of temporal coherence is complex and has a form completely analogous to the transverse case [Eq. (9.12b)]:

degree of temporal or longitudinal coherence for nearly monochromatic radiation

$$\gamma_\|(\tau) = \frac{\overline{\psi(t)\psi^*(t+\tau)}}{\overline{|\psi|^2}} \quad \text{for nearly monochromatic radiation.} \tag{9.15}$$

Here the average is over sufficiently long times t for the averaged value to settle down to an unchanging value.

When studying temporal coherence, one often wishes to deal with waves that contain a wide range of frequencies—such as the nearly Planckian (blackbody) cosmic microwave radiation emerging from the very early universe (Ex. 9.6). In this case, one should not factor any $e^{-i\omega_o t}$ out of the field Ψ, and one gains nothing by regarding $\Psi(t)$ as complex, so the *temporal coherence*

degree of temporal coherence for broadband radiation

$$\gamma_\|(\tau) = \frac{\overline{\Psi(t)\Psi(t+\tau)}}{\overline{\Psi^2}} \quad \text{for real } \Psi \text{ and broadband radiation} \tag{9.16}$$

is also real. We use this real $\gamma_\|$ throughout this subsection and the next. It obviously is the correlation function of Ψ [Eq. (6.19)] renormalized so $\gamma_\|(0) = 1$.

As τ is increased, $\gamma_\|$ typically remains near unity until some critical value τ_c is reached, and then it begins to fall off toward zero. The critical value τ_c, the longest time over which the field is strongly coherent, is the coherence time, of which we have already spoken: If the wave is roughly monochromatic, so $\Psi(t) \propto \cos[\omega_o t + \delta\varphi(t)]$, with ω_o fixed and the phase $\delta\varphi$ randomly varying in time, then it should be clear that the mean time for $\delta\varphi$ to change by an amount of order unity is the coherence time τ_c at which $\gamma_\|$ begins to fall significantly.

The uncertainty principle dictates that a field with coherence time τ_c, when Fourier analyzed in time, must contain significant power over a bandwidth $\Delta f = \Delta\omega/(2\pi) \sim 1/\tau_c$. Correspondingly, if we define the field's *longitudinal coherence length* by

longitudinal coherence
length

$$\boxed{l_\| \equiv c\tau_c,}$$

(9.17)

then $l_\|$ for broadband radiation will be only a few times the peak wavelength, but for a narrow spectral line of width $\Delta\lambda$, it will be $\lambda^2/\Delta\lambda$.

These relations between the coherence time or longitudinal coherence length and the field's spectral energy flux are order-of-magnitude consequences not only of the uncertainty relation, but also of the temporal analog of the van Cittert-Zernike theorem. That analog is just the Wiener-Khintchine theorem in disguise, and it can be derived by the same methods as we used in the transverse spatial domain. In that theorem the degree of lateral coherence $\gamma_\perp$ is replaced by the degree of temporal coherence $\gamma_\|$, and the angular intensity distribution $I(\alpha)$ (distribution of energy over angle) is replaced by the field's spectral energy flux $F_\omega(\omega)$ (the energy crossing a unit area per unit time and per unit angular frequency ω)—which is also called its *spectrum*.[5] The theorem takes the explicit form

spectrum or spectral
energy flux $F_\omega(\omega)$

$$\boxed{\begin{array}{c} \gamma_\|(\tau) = \dfrac{\int_{-\infty}^{\infty} d\omega\, F_\omega(\omega) e^{i\omega\tau}}{F_S} = \dfrac{2\int_0^{\infty} d\omega\, F_\omega(\omega) \cos\omega\tau}{F_S} \\ \text{for real } \Psi(t), \text{ valid for broadband radiation} \end{array}}$$

(9.18a)

temporal analog of van
Cittert-Zernike theorem

and

$$\boxed{F_\omega(\omega) = F_S \int_{-\infty}^{\infty} \frac{d\tau}{2\pi} \gamma_\|(\tau) e^{-i\omega\tau} = 2F_S \int_0^{\infty} \frac{d\tau}{2\pi} \gamma_\|(\tau) \cos\omega\tau.}$$

(9.18b)

[Here the normalization of our Fourier transform and the sign of its exponential are those conventionally used in optics, and differ from those used in the theory of random processes (Chap. 6). Also, because we have chosen Ψ to be real, $F_\omega(-\omega) = F_\omega(+\omega)$ and $\gamma_\|(-\tau) = \gamma_\|(+\tau)$.] One can measure $\gamma_\|$ by combining the radiation from two points displaced longitudinally to produce interference fringes just as we did when measuring spatial coherence. This type of interference is sometimes called *interference by division of the amplitude,* in contrast with "interference by division of the wavefront" for a Young's-slit-type measurement of lateral spatial coherence (next-to-last paragraph of Sec. 9.2.1).

interference by division of
the amplitude

---

5. Note that the spectral energy flux (spectrum) is simply related to the spectral density of the field: if the field Ψ is so normalized that the energy density is $U = \beta\,\overline{\Psi_{,t}\Psi_{,t}}$ with β some constant, then $F_\omega(\omega) = \beta c\omega^2/(2\pi)S_\Psi(f)$, with $f = \omega/(2\pi)$.

Exercise 9.5 *Problem: Longitudinal Coherence of Radio Waves*

An FM radio station has a carrier frequency of 91.3 MHz and transmits heavy metal rock music in frequency-modulated side bands of the carrier. Estimate the coherence length of the radiation.

9.2.7

Michelson interferometer

9.2.7 Michelson Interferometer and Fourier-Transform Spectroscopy

The classic instrument for measuring the degree of longitudinal coherence is the Michelson interferometer kf Fig. 9.3 (not to be confused with the Michelson stellar interferometer). In the simplest version, incident light (e.g., in the form of a Gaussian beam; Sec. 8.5.5) is split by a beam splitter into two beams, which are reflected off different plane mirrors and then recombined. The relative positions of the mirrors are adjustable so that the two light paths can have slightly different lengths. (An early version of this instrument was used in the famous Michelson-Morley experiment.) There are two ways to view the fringes. One way is to tilt one of the reflecting mirrors slightly so there is a range of path lengths in one of the arms. Light and dark interference bands (fringes) can then be seen across the circular cross section of the recombined beam. The second method is conceptually more direct but requires aligning the mirrors sufficiently accurately so the phase fronts of the two beams are parallel after recombination and the recombined beam has no banded structure. The end mirror in one arm of the interferometer is then slowly moved backward or forward, and as it moves, the recombined light slowly changes from dark to light to dark and so on.

It is interesting to interpret this second method in terms of the Doppler shift. One beam of light undergoes a Doppler shift on reflection off the moving mirror. There is then a beat wave produced when it is recombined with the unshifted radiation of the other beam.

Whichever method is used (tilted mirror or longitudinal motion of mirror), the visibility $\gamma_{\parallel}$ of the interference fringes measures the beam's degree of longitudinal coherence, which is related to the spectral energy flux (spectrum) F_ω by Eqs. (9.18).

Let us give an example. Suppose we observe a spectral line with rest angular frequency ω_0 that is broadened by random thermal motions of the emitting atoms. Then the line profile is

$$F_\omega \propto \exp\left(-\frac{(\omega_0 - \omega)^2}{2(\Delta\omega)^2}\right). \tag{9.19a}$$

The width of the line is given by the formula for the Doppler shift,

$$\Delta\omega \sim \omega_0 [k_B T/(mc^2)]^{1/2},$$

where T is the temperature of the emitting atoms, and m is their mass. (We ignore other sources of line broadening, e.g., natural broadening and pressure broadening,

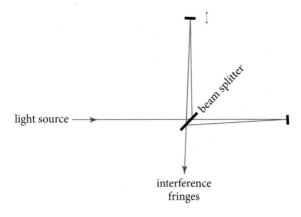

light source →

beam splitter

interference
fringes

FIGURE 9.3 Michelson interferometer.

which actually dominate under normal conditions.) For example, for hydrogen at $T = 10^3$ K, the Doppler-broadened line width is $\Delta\omega \sim 10^{-5}\omega_0$.

By Fourier transforming this line profile, using the well-known result that the Fourier transform of a Gaussian is another Gaussian and invoking the fundamental relations (9.18) between the spectrum and temporal coherence, we obtain

$$\gamma_\parallel(\tau) = \exp\left(-\frac{\tau^2(\Delta\omega)^2}{2}\right)\cos\omega_0\tau. \tag{9.19b}$$

If we had used the nearly monochromatic formalism with the field written as $\Psi = \psi(t)e^{-i\omega_0 t}$, then we would have obtained

$$\gamma_\parallel(\tau) = \exp\left(-\frac{\tau^2(\Delta\omega)^2}{2}\right)e^{i\omega_0\tau}, \tag{9.19c}$$

the real part of which is our broadband formalism's $\gamma_\parallel$. In either case, $\gamma_\parallel$ oscillates with angular frequency ω_0, and the amplitude of this oscillation is the *fringe visibility* V: **fringe visibility**

$$V = \exp\left(-\frac{\tau^2(\Delta\omega)^2}{2}\right). \tag{9.19d}$$

The variation $V(\tau)$ of this visibility with lag time τ is sometimes called an *interfero-*   **interferogram**
gram. For time lags $\tau \ll (\Delta\omega)^{-1}$, the line appears to be monochromatic, and fringes with unit visibility should be seen. However, for lags $\tau \gtrsim (\Delta\omega)^{-1}$, the fringe visibility will decrease exponentially with increasing τ^2. In our Doppler-broadened hydrogen-line example with $\Delta\omega \sim 10^{-5}\omega_0$, the rest angular frequency is $\omega_0 \sim 3 \times 10^{15}$ rad s^{-1}, so the longitudinal coherence length is $l_\parallel = c\tau_c \sim 10$ mm. No fringes will be seen when the radiation is combined from points separated by much more than this distance.

This procedure is an example of *Fourier transform spectroscopy,* in which, by   **Fourier transform**
measuring the degree of temporal coherence $\gamma_\parallel(\tau)$ and then Fourier transforming   **spectroscopy**

it [Eq. (9.18)], one infers the shape of the radiation's spectrum, or in this case, the width of a specific spectral line.

When (as in Ex. 9.6) the waves are very broad band, the degree of longitudinal coherence $\gamma_\parallel(\tau)$ will not have the form of a sinusoidal oscillation (regular fringes) with slowly varying amplitude (visibility). Nevertheless, the broadband van Cittert-Zernike theorem [Eqs. (9.18)] still guarantees that the spectrum (spectral energy flux) will be the Fourier transform of the coherence $\gamma_\parallel(\tau)$, which can be measured by a Michelson interferometer.

EXERCISES

Exercise 9.6 *Problem: COBE Measurement of the Cosmic Microwave Background Radiation*

An example of a Michelson interferometer is the Far Infrared Absolute Spectro-photometer (FIRAS) carried by the Cosmic Background Explorer satellite (COBE). COBE studied the spectrum and anisotropies of the cosmic microwave background radiation (CMB) that emerged from the very early, hot phase of our universe's expansion (Sec. 28.3.3). One of the goals of the COBE mission was to see whether the CMB spectrum really has the shape of 2.7 K blackbody (Planckian) radiation, or if it is highly distorted, as some measurements made on rocket flights had suggested. COBE's spectrophotometer used Fourier transform spectroscopy to meet this goal: it compared accurately the degree of longitudinal coherence $\gamma_\parallel$ of the CMB radiation with that of a calibrated source on board the spacecraft, which was known to be a blackbody at about 2.7 K. The comparison was made by alternately feeding radiation from the microwave background and radiation from the calibrated source into the same Michelson interferometer and comparing their fringe spacings. The result (Mather et al., 1994) was that the background radiation has a spectrum that is Planckian with temperature 2.726 ± 0.010 K over the wavelength range 0.5–5.0 mm, in agreement with simple cosmological theory that we shall explore in the last chapter of this book.

(a) Suppose that the CMB had had a Wien spectrum

$$F_\omega \propto \omega^3 \exp[-\hbar\omega/(k_B T)].$$

Show that the visibility of the fringes would have been

$$V = |\gamma_\parallel| \propto \frac{|s^4 - 6s_0^2 s^2 + s_0^4|}{(s^2 + s_0^2)^4} \tag{9.20}$$

where $s = c\tau$ is longitudinal distance, and calculate a numerical value for s_0.

(b) Compute the interferogram $V(\tau)$ for a Planck function either analytically (perhaps with the help of a computer) or numerically using a fast Fourier transform. Compare graphically the interferogram for the Wien and Planck spectra.

9.2.8 Degree of Coherence; Relation to Theory of Random Processes

Having separately discussed spatial and temporal coherence, we now can easily perform a final generalization and define the full degree of coherence of the radiation field between two points separated both laterally by a vector $\mathbf{a}$ and longitudinally by a distance s (or equivalently, by a time $\tau = s/c$). If we restrict ourselves to nearly monochromatic waves and use the complex formalism so the waves are written as $\Psi = e^{i(kz - \omega_o t)} \psi(\mathbf{x}, t)$ [Eq. (9.12a)], then we have

full (3-dimensional) degree of coherence

$$\gamma_{12}(k\mathbf{a}, \tau) \equiv \frac{\overline{\psi(\mathbf{x}_1, t)\psi^*(\mathbf{x}_1 + \mathbf{a}, t + \tau)}}{\left[\overline{|\psi(\mathbf{x}_1, t)|^2}\ \overline{|\psi(\mathbf{x}_1 + \mathbf{a}, t)|^2}\right]^{1/2}} = \frac{\overline{\psi(\mathbf{x}_1, t)\psi^*(\mathbf{x}_1 + \mathbf{a}, t + \tau)}}{\overline{|\psi|^2}}. \quad (9.21)$$

In the denominator of the second expression we have used the fact that, because the source is far away, $\overline{|\psi|^2}$ is independent of the spatial location at which it is evaluated, in the region of interest. Consistent with the definition (9.21), we can define a *volume of coherence* $\mathcal{V}_c$ as the product of the longitudinal coherence length $l_\parallel = c\tau_c$ and the square of the transverse coherence length $l_\perp^2$: $\mathcal{V}_c = l_\perp^2 c\tau_c$.

volume of coherence

The 3-dimensional version of the van Cittert-Zernike theorem relates the complex degree of coherence (9.21) to the radiation's *specific intensity*, $I_\omega(\boldsymbol{\alpha}, \omega)$, also called its spectral intensity (i.e., the energy crossing a unit area per unit time per unit solid angle and per unit angular frequency, or energy "per unit everything"). (Since the frequency ν and the angular frequency ω are related by $\omega = 2\pi\nu$, the specific intensity I_ω of this chapter and that I_ν of Chap. 3 are related by $I_\nu = 2\pi I_\omega$.) The 3-dimensional van Cittert-Zernike theorem states that

specific intensity

$$\gamma_{12}(k\mathbf{a}, \tau) = \frac{\int d\Omega_\alpha d\omega I_\omega(\boldsymbol{\alpha}, \omega) e^{i(k\mathbf{a}\cdot\boldsymbol{\alpha} + \omega\tau)}}{F_S}, \quad (9.22a)$$

and

3-dimensional van Cittert-Zernike theorem

$$I_\omega(\boldsymbol{\alpha}, \omega) = F_S \int \frac{d\tau d^2 k\mathbf{a}}{(2\pi)^3} \gamma_{12}(k\mathbf{a}, \tau) e^{-i(k\mathbf{a}\cdot\boldsymbol{\alpha} + \omega\tau)}. \quad (9.22b)$$

There obviously must be an intimate relationship between the theory of random processes, as developed in Chap. 6, and the theory of a wave's coherence, as we have developed it in Sec. 9.2. That relationship is explained in Ex. 9.8.

EXERCISES

Exercise 9.7 *Problem: Decomposition of Degree of Coherence*
We have defined the degree of coherence $\gamma_{12}(\mathbf{a}, \tau)$ for two points in the radiation field separated laterally by a distance $\mathbf{a}$ and longitudinally by a time τ. Under what conditions will this be given by the product of the spatial and temporal degrees of coherence?

$$\gamma_{12}(\mathbf{a}, \tau) = \gamma_\perp(\mathbf{a})\gamma_\parallel(\tau). \quad (9.23)$$

Exercise 9.8 **Example: Complex Random Processes and the van Cittert-Zernike Theorem*

In Chap. 6 we developed the theory of real-valued random processes that vary randomly with time t (i.e., that are defined on a 1-dimensional space in which t is a coordinate). Here we generalize a few elements of that theory to a complex-valued random process $\Phi(\mathbf{x})$ defined on a (Euclidean) space with n dimensions. We assume the process to be stationary and to have vanishing mean (cf. Chap. 6 for definitions). For $\Phi(\mathbf{x})$ we define a complex-valued correlation function by

$$C_\Phi(\boldsymbol{\xi}) \equiv \overline{\Phi(\mathbf{x})\Phi^*(\mathbf{x} + \boldsymbol{\xi})} \tag{9.24a}$$

(where $*$ denotes complex conjugation) and a real-valued spectral density by

$$S_\Phi(\mathbf{k}) = \lim_{L\to\infty} \frac{1}{L^n}|\tilde{\Phi}_L(\mathbf{k})|^2. \tag{9.24b}$$

Here Φ_L is Φ confined to a box of side L (i.e., set to zero outside that box), and the tilde denotes a Fourier transform defined using the conventions of Chap. 6:

$$\tilde{\Phi}_L(\mathbf{k}) = \int \Phi_L(\mathbf{x})e^{-i\mathbf{k}\cdot\mathbf{x}}d^n x, \qquad \Phi_L(\mathbf{x}) = \int \tilde{\Phi}_L(\mathbf{k})e^{+i\mathbf{k}\cdot\mathbf{x}}\frac{d^n k}{(2\pi)^n}. \tag{9.25}$$

Because Φ is complex rather than real, $C_\Phi(\boldsymbol{\xi})$ is complex; and as we shall see below, its complexity implies that [although $S_\Phi(\mathbf{k})$ is real], $S_\Phi(-\mathbf{k}) \neq S_\Phi(\mathbf{k})$. This fact prevents us from folding negative $\mathbf{k}$ into positive $\mathbf{k}$ and thereby making $S_\Phi(\mathbf{k})$ into a "single-sided" spectral density as we did for real random processes in Chap. 6. In this complex case we must distinguish $-\mathbf{k}$ from $+\mathbf{k}$ and similarly $-\boldsymbol{\xi}$ from $+\boldsymbol{\xi}$.

(a) The complex Wiener-Khintchine theorem [analog of Eq. (6.29)] states that

$$S_\Phi(\mathbf{k}) = \int C_\Phi(\boldsymbol{\xi})e^{+i\mathbf{k}\cdot\boldsymbol{\xi}}d^n\xi, \tag{9.26a}$$

$$C_\Phi(\boldsymbol{\xi}) = \int S_\Phi(\mathbf{k})e^{-i\mathbf{k}\cdot\boldsymbol{\xi}}\frac{d^n k}{(2\pi)^n}. \tag{9.26b}$$

Derive these relations. [Hint: Use Parseval's theorem in the form $\int A(\mathbf{x})B^*(\mathbf{x})d^n x = \int \tilde{A}(\mathbf{k})\tilde{B}^*(\mathbf{k})d^n k/(2\pi)^n$ with $A(\mathbf{x}) = \Phi_L(\mathbf{x})$ and $B(\mathbf{x}) = \Phi_L(\mathbf{x} + \boldsymbol{\xi})$, and then take the limit as $L \to \infty$.] Because $S_\Phi(\mathbf{k})$ is real, this Wiener-Khintchine theorem implies that $C_\Phi(-\boldsymbol{\xi}) = C_\Phi^*(\boldsymbol{\xi})$. Show that this is so directly from the definition (9.24a) of $C_\Phi(\boldsymbol{\xi})$. Because $C_\Phi(\boldsymbol{\xi})$ is complex, the Wiener-Khintchine theorem implies that $S_\Phi(\mathbf{k}) \neq S_\Phi(-\mathbf{k})$.

(b) Let $\psi(\mathbf{x}, t)$ be the complex-valued wave field defined in Eq. (9.12a), and restrict $\mathbf{x}$ to range only over the two transverse dimensions so ψ is defined on a 3-dimensional space. Define $\Phi(\mathbf{x}, t) \equiv \psi(\mathbf{x}, t)/\left[\overline{|\psi(\mathbf{x}, t)|^2}\right]^{1/2}$. Show that

$$C_\Phi(\mathbf{a}, \tau) = \gamma_{12}(k\mathbf{a}, \tau), \quad S_\Phi(-\boldsymbol{\alpha}k, -\omega) = \text{const} \times \frac{I_\omega(\boldsymbol{\alpha}, \omega)}{F_S}, \tag{9.27}$$

and that the complex Wiener-Khintchine theorem (9.26) is the van Cittert-Zernike theorem (9.22). (Note: The minus signs in S_Φ result from the difference in Fourier transform conventions between the theory of random processes [Eq. (9.25) and Chap. 6] and the theory of optical coherence [this chapter].) Evaluate the constant in Eq. (9.27).

9.3 Radio Telescopes

The interferometry technique pioneered by Michelson for measuring the angular sizes of stars at visual wavelengths has been applied to great effect in radio astronomy.

A modern radio telescope is a large surface that reflects radio waves onto a "feed," where the waves' fluctuating electric field creates a tiny electric voltage that subsequently can be amplified and measured electronically. Modern radio telescopes have diameters D that range from ~ 10 m to the ~ 300 m of the Arecibo telescope in Puerto Rico. A typical observing wavelength might be $\lambda \sim 6$ cm. This implies an angular resolution $\theta_A \sim \lambda/D \sim 2 \, \text{arcmin}(\lambda/6 \, \text{cm})(D/100 \, \text{m})^{-1}$ [Eq. (8.18) and subsequent discussion]. However, many of the most interesting cosmic sources are much smaller than this. To achieve much better angular resolution, the technique of radio interferometry was pioneered in the 1950s and has been steadily developing since then.[6]

radio telescope interferometry

9.3.1 Two-Element Radio Interferometer

If we have two radio telescopes, then we can think of them as two Young's slits, and we can link them using a combination of waveguides and electric cables, as shown in Fig. 9.4. When they are both pointed at a source, they both measure the electric field in radio waves from that source. We combine their signals by narrowband filtering their voltages to make them nearly monochromatic and then either adding the filtered voltages and measuring the power, or multiplying the two voltages directly. In either case a measurement of the degree of coherence [Eq. (9.10)] can be achieved. [If the source is not vertically above the two telescopes, one obtains some nonlateral component of the full degree of coherence $\gamma_{12}(\mathbf{a}, \tau)$. However, by introducing a time delay into one of the signals, as in Fig. 9.4, one can measure the degree of lateral coherence $\gamma_\perp(\mathbf{a})$, which is what the astronomer usually needs.]

The objective is usually to produce an image of the radio waves' source. This is achieved by Fourier inverting the lateral degree of coherence $\gamma_\perp(\mathbf{a})$ [Eq. (9.13b)], which therefore must be measured for a variety of values of the relative separation vector $\mathbf{a}$ of the telescopes perpendicular to the source's direction. As Earth rotates, the separation vector will trace out half an ellipse in the 2-dimensional $\mathbf{a}$ plane every 12 hours. [The source intensity is a real quantity, so we can use Eq. (9.13a) to deduce

---

6. This type of interferometry is also developing fast at optical wavelengths. This was not possible until optical technology became good enough to monitor the phase of light as well as its amplitude, at separate locations, and then produce interference.

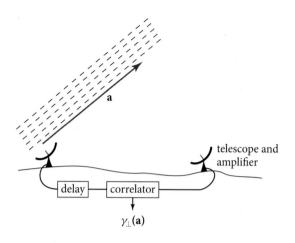

FIGURE 9.4 Two-element radio interferometer.

that $\gamma_\perp(-\mathbf{a}) = \gamma_\perp^*(\mathbf{a})$, which gives the other half of the ellipse.] By changing the spacing between the two telescopes daily and collecting data for a number of days, the degree of coherence can be well sampled. This technique[7] is known as *Earth-rotation aperture synthesis*, because the telescopes are being made to have the angular resolution of a giant telescope as big as their maximum separation, with the aid of Earth's rotation. They do not, of course, have the sensitivity of this giant telescope.

aperture synthesis

9.3.2 Multiple-Element Radio Interferometers

In practice, a modern radio interferometer has many more than two telescopes. For example, the Karl G. Jansky Very Large Array (JVLA) in New Mexico (USA) has 27 individual telescopes arranged in a Y pattern and operating simultaneously, with a maximum baseline of 36 km and a minimum observing wavelength of 7 mm. The degree of coherence can thus be measured simultaneously over $27 \times 26/2 = 351$ different relative separations. The results of these measurements can then be interpolated to give values of $\gamma_\perp(\mathbf{a})$ on a regular grid of points (usually $2^N \times 2^N$ for some integer N). This is then suitable for applying the fast Fourier transform algorithm to infer the source structure $I(\boldsymbol{\alpha})$.

JVLA

The Atacama Large Millimeter Array (ALMA) being constructed in Chile is already (2016) operational. It comprises 66 telescopes observing with wavelengths between 0.3 and 9.6 mm and baselines as long as 16 km. Future ambitions at longer radio wavelengths are centered on the proposed Square Kilometer Array (SKA) to be built in South Africa and Australia comprising thousands of dishes with a combined collecting area of a square kilometer.

ALMA

7. For which Martin Ryle was awarded the Nobel Prize.

9.3.3 Closure Phase

Among the many technical complications of interferometry is one that brings out an interesting point about Fourier methods. It is usually much easier to measure the modulus than the phase of the complex degree of coherence. This is partly because **phase errors** it is hard to introduce the necessary delays in the electronics accurately enough to know where the zero of the fringe pattern should be located and partly because unknown, fluctuating phase delays are introduced into the phase of the field as the wave propagates through the upper atmosphere and ionosphere. [This is a radio variant of the problem of "seeing" for optical telescopes (cf. Box 9.2), and it also plagues the Michelson stellar interferometer.] It might therefore be thought that we would have to make do with just the modulus of the degree of coherence (i.e., the fringe visibility) to perform the Fourier inversion for the source structure. This is not so.

Consider a three-element interferometer measuring fields ψ_1, ψ_2, and ψ_3, and suppose that at each telescope there are unknown phase errors, $\delta\varphi_1$, $\delta\varphi_2$, and $\delta\varphi_3$ (Fig. 9.5). For baseline $\mathbf{a}_{12}$, we measure the degree of coherence $\gamma_{\perp 12} \propto \overline{\psi_1\psi_2^*}$, a complex number with phase $\Phi_{12} = \varphi_{12} + \delta\varphi_1 - \delta\varphi_2$, where φ_{12} is the phase of $\gamma_{\perp 12}$ in the absence of phase errors. If we also measure the degrees of coherence for the other two pairs of telescopes in the triangle and derive their phases Φ_{23} and Φ_{31}, we can then calculate the quantity

$$C_{123} = \Phi_{12} + \Phi_{23} + \Phi_{31}$$
$$= \varphi_{12} + \varphi_{23} + \varphi_{31}, \tag{9.28}$$

from which the phase errors cancel out.

The quantity C_{123}, known as the *closure phase,* can be measured with high accu- **closure phase** racy. In the JVLA, there are $27 \times 26 \times 25/6 = 2{,}925$ such closure phases, and they can all be measured with considerable redundancy. Although absolute phase information cannot be recovered, 93% of the telescopes' relative phases can be inferred in this manner and used to construct an image far superior to what could be achieved without any phase information.

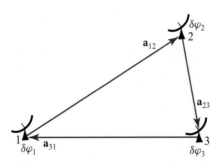

FIGURE 9.5 Closure-phase measurement using a triangle of telescopes.

9.3.4 Angular Resolution

When the telescope spacings are well sampled and the source is bright enough to carry out these image-processing techniques, an interferometer can have an angular resolving power approaching that of an equivalent filled aperture as large as the maximum telescope spacing. For the JVLA the best angular resolution is 50 milliarcsec and at ALMA it will be as fine as 4 milliarcsec, thousands of times better than single dishes.

Even greater angular resolution is achieved with a technique known as very long baseline interferometry (VLBI). Here the telescopes can be located on different continents and instead of linking them directly, the oscillating field amplitudes $\psi(t)$ are stored electronically. Then they are combined digitally long after the observation to compute the complex degree of coherence and thence the source structure $I(\boldsymbol{\alpha})$. In this way angular resolutions more than 300 times better than those achievable by the JVLA have been obtained. Structure smaller than a milliarcsec, corresponding to a few light-years at cosmological distances, can be measured in this manner. In an impressive recent observation (The Event Horizon Telescope Collaboration, 2019), an image has been made, at a wavelength of 1.3 mm, of the region around the 6.5 billion-solar-mass black hole in the nucleus of the nearby galaxy M87. The angular resolution is roughly 20 microarcsec $\sim 5\,GM/Dc^2$, where D is the distance. Emitting plasma is seen orbiting the black hole, which casts a dark shadow.

EXERCISES

Exercise 9.9 *Example: Radio Interferometry from Space*
The longest radio-telescope separation available in 2016 is that between telescopes on Earth's surface and a 10-m diameter radio telescope in the Russian RadioAstron satellite, which was launched into a highly elliptical orbit around Earth in summer 2011, with perigee $\sim$10,000 km (1.6 Earth radii) and apogee $\sim$350,000 km (55 Earth radii).

(a) Radio astronomers conventionally describe the specific intensity $I_\omega(\boldsymbol{\alpha}, \omega)$ of a source in terms of its brightness temperature. This is the temperature $T_b(\omega)$ that a blackbody would have to emit, in the Rayleigh-Jeans (low-frequency) end of its spectrum, to produce the same specific intensity as the source. Show that for a single (linear or circular) polarization, if the solid angle subtended by a source is $\Delta\Omega$ and the spectral energy flux measured from the source is $F_\omega \equiv \int I_\omega d\Omega = I_\omega \Delta\Omega$, then the brightness temperature is

$$T_b = \frac{(2\pi)^3 c^2 I_\omega}{k_B \omega^2} = \frac{(2\pi)^3 c^2 F_\omega}{k_B \omega^2 \Delta\Omega}, \tag{9.29}$$

where k_B is Boltzmann's constant.

(b) The brightest quasars emit radio spectral fluxes of about $F_\omega = 10^{-25}\,\mathrm{W\,m^{-2}\,Hz^{-1}}$, independent of frequency. The smaller such a quasar is, the larger will be its brightness temperature. Thus, one can characterize the smallest sources that a radio-telescope system can resolve by the highest brightness temperatures it can measure. Show that the maximum brightness temperature measurable by the Earth-to-orbit RadioAstron interferometer is independent of the frequency at which the observation is made, and estimate its numerical value.

9.4 Etalons and Fabry-Perot Interferometers

We have shown how a Michelson interferometer (Fig. 9.3) can be used as a Fourier-transform spectrometer: one measures the complex fringe visibility as a function of the two arms' optical path difference and then takes the visibility's Fourier transform to obtain the spectrum of the radiation. The inverse process is also powerful. One can drive a Michelson interferometer with radiation with a known, steady spectrum (usually close to monochromatic), and look for time variations of the positions of its fringes caused by changes in the relative optical path lengths of the interferometer's two arms. This was the philosophy of the famous Michelson-Morley experiment to search for ether drift, and it is also the underlying principle of a laser interferometer ("interferometric") gravitational-wave detector.

To reach the sensitivity required for gravitational-wave detection, one must modify the Michelson interferometer by making the light travel back and forth in each arm many times, thereby amplifying the phase shift caused by changes in the arm lengths. This is achieved by converting each arm into a Fabry-Perot interferometer. In this section, we study Fabry-Perot interferometers and some of their other applications, and in Sec. 9.5, we explore their use in gravitational-wave detection.

9.4.1 Multiple-Beam Interferometry; Etalons

Fabry-Perot interferometry is based on trapping monochromatic light between two highly reflecting surfaces. To understand such trapping, let us consider the concrete situation where the reflecting surfaces are flat and parallel to each other, and the transparent medium between the surfaces has one index of refraction n, while the medium outside the surfaces has another index n' (Fig. 9.6). Such a device is sometimes called an *etalon*. One example is a glass slab in air ($n \simeq 1.5$, $n' \simeq 1$); another is a vacuum maintained between two glass mirrors ($n = 1$, $n' \simeq 1.5$). For concreteness, we discuss the slab case, though all our formulas are equally valid for a vacuum between mirrors or for any other etalon.

etalon

Suppose that a monochromatic plane wave (i.e., with parallel rays) with angular frequency ω is incident on one of the slab's reflecting surfaces, where it is partially reflected and partially transmitted with refraction. The transmitted wave will propagate through to the second surface, where it will be partially reflected and partially transmitted. The reflected portion will return to the first surface, where it too will be split, and so on (Fig. 9.6a). The resulting total fields in and outside the slab can be computed by summing the series of sequential reflections and transmissions (Ex. 9.12). Alternatively, they can be computed as follows.

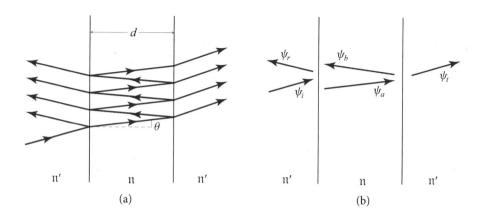

FIGURE 9.6 Multiple-beam interferometry using a type of Fabry-Perot etalon.

Assume, for pedagogical simplicity, that there is translational invariance along the slab (i.e., the slab and incoming wave are perfectly planar). Then the series, if summed, would lead to the five waves shown in Fig. 9.6b: an incident wave (ψ_i), a reflected wave (ψ_r), a transmitted wave (ψ_t), and two internal waves (ψ_a and ψ_b).

amplitude reflection and transmission coefficients

We introduce amplitude reflection and transmission coefficients, denoted $\mathfrak{r}$ and $\mathfrak{t}$, for waves incident on the slab surface from outside. Likewise, we introduce coefficients $\mathfrak{r}'$, $\mathfrak{t}'$ for waves incident on the slab from inside. These coefficients are functions of the angles of incidence and the light's polarization. They can be computed using electromagnetic theory (e.g., Hecht, 2017, Sec. 4.6.2), but this will not concern us here.

Armed with these definitions, we can express the reflected and transmitted waves at the first surface (location A in Fig. 9.7) in the form

$$\psi_r = \mathfrak{r}\psi_i + \mathfrak{t}'\psi_b,$$
$$\psi_a = \mathfrak{t}\psi_i + \mathfrak{r}'\psi_b, \tag{9.30a}$$

where ψ_i, ψ_a, ψ_b, and ψ_r are the values of ψ at A for waves impinging on or leaving the surface along the paths i, a, b, and r, respectively, depicted in Fig. 9.7. Simple geometry shows that the waves at the second surface are as depicted in Fig. 9.7. Correspondingly, the relationships between the ingoing and outgoing waves there are

$$\psi_b e^{-iks_1} = \mathfrak{r}'\psi_a e^{ik(s_1-s_2)},$$
$$\psi_t = \mathfrak{t}'\psi_a e^{iks_1}, \tag{9.30b}$$

where $k = n\omega/c$ is the wave number in the slab, and (as is shown in the figure) s_1 and s_2 are defined as

$$s_1 = d \sec\theta, \qquad s_2 = 2d \tan\theta \sin\theta, \tag{9.30c}$$

with d the thickness of the slab, and θ the angle that the wavefronts inside the slab make to the slab's faces.

In solving Eqs. (9.30) for the net transmitted and reflected waves ψ_t and ψ_r in terms of the incident wave ψ_i, we need *reciprocity relations* between the reflection

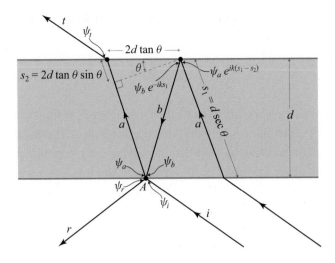

FIGURE 9.7 Construction for calculating the phase differences across the slab for the two internal waves in an etalon.

and transmission coefficients $\mathfrak{r}$ and $\mathfrak{t}$ for waves that hit the reflecting surfaces from one side, and those between $\mathfrak{r}'$ and $\mathfrak{t}'$ for waves from the other side. These reciprocity relations are analyzed quite generally in Ex. 9.10. To derive the reciprocity relations in our case of sharp boundaries between homogeneous media, consider the limit in which the slab thickness $d \rightarrow 0$. This is allowed because the wave equation is linear, and the solution for one surface can be superposed on that for the other surface. In this limit $s_1 = s_2 = 0$ and the slab must become transparent, so

$$\psi_r = 0, \qquad \psi_t = \psi_i. \tag{9.31}$$

Equations (9.30a), (9.30b), and (9.31) are then six homogeneous equations in the five wave amplitudes ψ_i, ψ_r, ψ_t, ψ_a, and ψ_b, from which we can extract the two desired reciprocity relations:

$$\boxed{\mathfrak{r}' = -\mathfrak{r}, \qquad \mathfrak{t}\mathfrak{t}' - \mathfrak{r}\mathfrak{r}' = 1.} \tag{9.32}$$

reciprocity relations

Since there is no mechanism to produce a phase shift as the waves propagate across a perfectly sharp boundary, it is reasonable to expect $\mathfrak{r}$, $\mathfrak{r}'$, $\mathfrak{t}$, and $\mathfrak{t}'$ all to be real, as indeed they are (Ex. 9.10). [If the interface has a finite thickness, it is possible to adjust the spatial origins on the two sides of the interface so as to make $\mathfrak{r}$, $\mathfrak{r}'$, $\mathfrak{t}$, and $\mathfrak{t}'$ all be real, leading to the reciprocity relations (9.32), but a price will be paid; see Ex. 9.10.]

Now return to the case of finite slab thickness. By solving Eqs. (9.30) for the reflected and transmitted fields and invoking the reciprocity relations (9.32), we obtain

$$\frac{\psi_r}{\psi_i} \equiv \mathfrak{r}_e = \frac{\mathfrak{r}(1 - e^{i\varphi})}{1 - \mathfrak{r}^2 e^{i\varphi}}, \qquad \frac{\psi_t}{\psi_i} \equiv \mathfrak{t}_e = \frac{(1 - \mathfrak{r}^2)e^{i\varphi/(2\cos^2\theta)}}{1 - \mathfrak{r}^2 e^{i\varphi}}. \tag{9.33a}$$

Here $\mathfrak{r}_e$ and $\mathfrak{t}_e$ are the etalon's reflection and transmission coefficients, and $\varphi = k(2s_1 - s_2)$, which reduces to

$$\varphi = 2\mathfrak{n}\omega d \cos \theta / c, \qquad (9.33b)$$

is the light's round-trip phase shift (along path a and then b) inside the etalon, relative to the phase of the incoming light that it meets at location A. If φ is a multiple of 2π, the round-trip light will superpose coherently on the new, incoming light.

We are particularly interested in the *reflectivity and transmissivity* for the energy flux—the coefficients that tell us what fraction of the total flux (and therefore also the total power) incident on the etalon is reflected by it and what fraction emerges from its other side:

etalon's reflectivity and transmissivity

$$R = |\mathfrak{r}_e|^2 = \frac{|\psi_r|^2}{|\psi_i|^2} = \frac{2\mathfrak{r}^2(1 - \cos \varphi)}{1 - 2\mathfrak{r}^2 \cos \varphi + \mathfrak{r}^4}, \qquad T = |\mathfrak{t}_e|^2 = \frac{|\psi_t|^2}{|\psi_i|^2} = \frac{(1 - \mathfrak{r}^2)^2}{1 - 2\mathfrak{r}^2 \cos \varphi + \mathfrak{r}^4}.$$

$$(9.33c)$$

From these expressions, we see that

energy conservation

$$\boxed{R + T = 1,} \qquad (9.33d)$$

which says that the energy flux reflected from the slab plus that transmitted is equal to that impinging on the slab (energy conservation). It is actually the reciprocity relations (9.32) for the amplitude reflection and transmission coefficients that enforce this energy conservation. If they had contained a provision for absorption or scattering of light in the interfaces, $R + T$ would have been less than one.

We discuss the etalon's reflectivity and transmissivity, Eq. (9.33c), at length in Sec. 9.4.2. But first, in a set of example exercises, we clarify some important issues related to the above analysis.

EXERCISES

Exercise 9.10 *Example: Reciprocity Relations for a Locally Planar Optical Device*
Modern mirrors, etalons, beam splitters, and other optical devices are generally made of glass or fused silica (quartz) and have dielectric coatings on their surfaces. The coatings consist of alternating layers of materials with different dielectric constants, so the index of refraction $\mathfrak{n}$ varies periodically. If, for example, the period of $\mathfrak{n}$'s variations is half a wavelength of the radiation, then waves reflected from successive dielectric layers build up coherently, producing a large net reflection coefficient; the result is a highly reflecting mirror.

In this exercise, we use a method due to Stokes to derive the reciprocity relations for devices with dielectric coatings, and in fact for much more general devices. Specifically, our derivation will be valid for locally plane-fronted, monochromatic waves

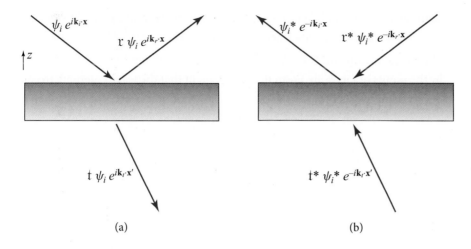

FIGURE 9.8 Construction for deriving reciprocity relations for amplitude transmission and reflection coefficients.

impinging on an arbitrary, locally planar, lossless optical device.[8] The device could be a mirror, a surface with an antireflection coating (Ex. 9.13 below), an etalon, or any sequence of such objects with locally parallel surfaces.

Let a plane, monochromatic wave $\psi_i e^{i\mathbf{k}_i \cdot \mathbf{x}} e^{-i\omega t}$ impinge on the optical device from above, and orient the device so its normal is in the z direction and it is translation invariant in the x and y directions; see Fig. 9.8a. Then the reflected and transmitted waves are as shown in the figure. Because the medium below the device can have a different index of refraction from that above, the waves' propagation direction below may be different from that above, as shown. For reasons explained in part (e), we denote position below the device by $\mathbf{x}'$ and position above the device by $\mathbf{x}$. Some arbitrary choice has been made for the locations of the vertical origins $z = 0$ and $z' = 0$ on the two sides of the device.

(a) Consider a thought experiment in which the waves of Fig. 9.8a are time-reversed, so they impinge on the device from the original reflection and transmission directions and emerge toward the original input direction, as shown in Fig. 9.8b. If the device had been lossy, the time-reversed waves would not satisfy the field's wave equation; the absence of losses guarantees they do. Show that mathematically, the time reversal can be achieved by complex conjugating the spatial part of the waves while leaving the temporal part $e^{-i\omega t}$ unchanged. (Such phase conjugation can be achieved in practice using techniques of nonlinear optics, as we

8. By "locally" plane-fronted and planar, we mean that transverse variations are on scales sufficiently long compared to the wavelength of light that we can use the plane-wave analysis sketched here; for example, the spherical mirrors and Gaussian beams of an interferometric gravitational-wave detector (see Fig. 9.13 in Sec. 9.5) easily satisfy this requirement. By lossless we mean that there is no absorption or scattering of the light.

shall see in the next chapter.) Show, correspondingly, that the spatial part of the time-reversed waves is described by the formulas shown in Fig. 9.8b.

(b) Use the reflection and transmission coefficients to compute the waves produced by the inputs of Fig. 9.8b. From the requirement that the wave emerging from the device's upward side must have the form shown in the figure, conclude that

$$1 = \mathfrak{r}\mathfrak{r}^* + \mathfrak{t}'\mathfrak{t}^*. \tag{9.34a}$$

Similarly, from the requirement that no wave emerge from the device's downward side, conclude that

$$0 = \mathfrak{t}\mathfrak{r}^* + \mathfrak{t}^*\mathfrak{r}'. \tag{9.34b}$$

Eqs. (9.34) are the most general form of the reciprocity relations for lossless, planar devices.

(c) For a sharp interface between two homogeneous media, combine these general reciprocity relations with the ones derived in the text [Eqs. (9.32)] to show that $\mathfrak{t}$, $\mathfrak{t}'$, $\mathfrak{r}$, and $\mathfrak{r}'$ are all real (as was asserted in the text).

(d) For the etalon of Figs. 9.6 and 9.7, $\mathfrak{r}_e$ and $\mathfrak{t}_e$ are given by Eqs. (9.33a). What do the reciprocity relations tell us about the coefficients for light propagating in the opposite direction, $\mathfrak{r}'_e$ and $\mathfrak{t}'_e$?

(e) Show that for a general optical device, the reflection and transmission coefficients can all be made real by appropriate, independent *adjustments of the origins of the vertical coordinates* z (for points above the device) and z' (for points below the device). More specifically, show that by setting $z_{\text{new}} = z_{\text{old}} + \delta z$ and $z'_{\text{new}} = z'_{\text{old}} + \delta z'$ and choosing δz and $\delta z'$ appropriately, one can make $\mathfrak{t}$ and $\mathfrak{r}$ real. Show further that the reciprocity relations (9.34a) and (9.34b) then imply that $\mathfrak{t}'$ and $\mathfrak{r}'$ are also real. Finally, show that this adjustment of origins brings the real reciprocity relations into the same form (9.32) as for a sharp interface between two homogeneous media.

As attractive as it may be to have these coefficients real, one must keep in mind some disadvantages: (i) the displaced origins for z and z' in general will depend on frequency, and correspondingly, (ii) frequency-dependent information (most importantly, frequency-dependent phase shifts of the light) is lost by making the coefficients real. If the phase shifts depend only weakly on frequency over the band of interest (as is typically the case for the dielectric coating of a mirror face), then these disadvantages are unimportant and it is conventional to choose the coefficients real. If the phase shifts depend strongly on frequency over the band of interest [e.g., for the etalon of Eqs. (9.33a), when its two faces are highly reflecting and its round-trip phase φ is near a multiple of 2π], the disadvantages are severe. One then should leave the origins frequency independent, and correspondingly leave the device's $\mathfrak{r}$, $\mathfrak{r}'$, $\mathfrak{t}$, and $\mathfrak{t}'$ complex [as we have for the etalon in Eqs. (9.33a)].

Exercise 9.11 **Example: Transmission and Reflection Coefficients for an Interface between Dielectric Media*

Consider monochromatic electromagnetic waves that propagate from a medium with index of refraction n_1 into a medium with index of refraction n_2. Let z be a Cartesian coordinate perpendicular to the planar interface between the media.

(a) From the Helmholtz equation $[-\omega^2 + (c^2/n^2)\nabla^2]\psi = 0$, show that both ψ and $\psi_{,z}$ must be continuous across the interface.

(b) Using these continuity requirements, show that for light propagating orthogonal to the interface (z direction), the reflection and transmission coefficients, in going from medium 1 to medium 2, are

$$\boxed{r = \frac{n_1 - n_2}{n_1 + n_2}, \qquad t = \frac{2n_1}{n_1 + n_2}.} \qquad (9.35)$$

Notice that these r and t are both real.

(c) Use the reciprocity relations (9.34) to deduce the reflection and transmission coefficients r' and t' for a wave propagating in the opposite direction, from medium 2 to medium 1.

Exercise 9.12 **Example: Etalon's Light Fields Computed by Summing the Contributions from a Sequence of Round Trips*

Study the step-by-step buildup of the field inside an etalon and the etalon's transmitted field, when the input field is suddenly turned on. More specifically, carry out the following steps.

(a) When the wave first turns on, the transmitted field inside the etalon, at point A of Fig. 9.7, is $\psi_a = t\psi_i$, which is very small if the reflectivity is high so that $|t| \ll 1$. Show (with the aid of Fig. 9.7) that, after one round-trip-travel time in the etalon, the transmitted field at A is $\psi_a = t\psi_i + (r')^2 e^{i\varphi}t\psi_i$. Show that for high reflectivity and on resonance, the tiny transmitted field has doubled in amplitude and its energy flux has quadrupled.

(b) Compute the transmitted field ψ_a at A after more and more round trips, and watch it build up. Sum the series to obtain the steady-state field ψ_a. Explain the final, steady-state amplitude: why is it not infinite, and why, physically, does it have the value you have derived?

(c) Show that, at any time during this buildup, the field transmitted out the far side of the etalon is $\psi_t = t'\psi_a e^{iks_1}$ [Eq. (9.30b)]. What is the final, steady-state transmitted field? Your answer should be Eqs. (9.33a).

Exercise 9.13 **Example: Anti-Reflection Coating*

A common technique used to reduce the reflection at the surface of a lens is to coat it with a quarter wavelength of material with refractive index equal to the geometric mean of the refractive indices of air and glass.

(a) Show that this does indeed lead to perfect transmission of normally incident light.

(b) Roughly how thick must the layer be to avoid reflection of blue light? Estimate the energy-flux reflection coefficient for red light in this case.

[Hint: The amplitude reflection coefficients at an interface are given by Eqs. (9.35).]

Exercise 9.14 *Problem: Oil Slick*
When a thin layer of oil lies on top of water, one sometimes sees beautiful, multi-colored, irregular bands of light reflecting off the oil layer. Explain qualitatively what causes this.

9.4.2

9.4.2 Fabry-Perot Interferometer and Modes of a Fabry-Perot Cavity with Spherical Mirrors

When an etalon's two faces are highly reflecting (reflection coefficient $\mathfrak{r}$ near unity), we can think of them as mirrors, between which the light resonates. The etalon is then a special case of a *Fabry-Perot interferometer*. The general case is any device in which light resonates between two high-reflectivity mirrors. The mirrors need not be planar and need not have the same reflectivities, and the resonating light need not be plane fronted.

Fabry-Perot interferometer

A common example is the optical cavity of Fig. 7.9, formed by two mirrors that are segments of spheres, which we studied using geometric optics in Ex. 7.12. Because the phase fronts of a Gaussian beam (Sec. 8.5.5) are also spherical, such a beam can resonate in the optical cavity if (i) the beam's waist location and waist radius are adjusted so its phase-front radii of curvature, at the mirrors, are the same as the mirrors' radii of curvature, and (ii) the light's frequency is adjusted so a half-integral number of wavelengths fit perfectly inside the cavity. Box 9.3 gives details for the case where the two mirrors have identical radii of curvature. In that box we also learn that the Gaussian beams are not the only eigenmodes that can resonate inside such a cavity. Other, "higher-order" modes can also resonate. They have more complex transverse distributions of the light. There are two families of such modes: one with rectangular transverse light distributions, and the other with wedge-shaped, spoke-like light distributions.

For any Fabry-Perot interferometer with identical mirrors, driven by light with a transverse cross section that matches one of the interferometer's modes, one can study the interferometer's response to the driving light by the same kind of analysis as we used for an etalon in the previous section. And the result will be the same: the interferometer's reflected and transmitted light, at a given transverse location $\{x, y\}$, will be given by

$$
\frac{\psi_r}{\psi_i} \equiv \mathfrak{r}_{FP} = \frac{\mathfrak{r}(1 - e^{i\varphi})}{1 - \mathfrak{r}^2 e^{i\varphi}}, \qquad \frac{\psi_t}{\psi_i} \equiv \mathfrak{t}_{FP} = \frac{(1 - \mathfrak{r}^2)e^{i\varphi/2}}{1 - \mathfrak{r}^2 e^{i\varphi}} \tag{9.36}
$$

BOX 9.3. MODES OF A FABRY-PEROT CAVITY WITH SPHERICAL MIRRORS

Consider a Fabry-Perot cavity whose spherical mirrors have the same radius of curvature R and are separated by a distance L. Introduce (i) Cartesian coordinates with $z = 0$ at the cavity's center, and (ii) the same functions we used for Gaussian beams [Eqs. (8.40b)]:

$$z_0 = \frac{k\sigma_0^2}{2} = \frac{\pi\sigma_0^2}{\lambda}, \qquad \sigma_z = \sigma_0(1 + z^2/z_0^2)^{1/2}, \qquad R_z = z(1 + z_0^2/z^2), \quad (1)$$

with k the wave number and σ_0 a measure of the transverse size of the beam at the cavity's center. Then it is straightforward to verify that the following functions (i) satisfy the Helmholtz equation, (ii) are orthonormal when integrated over their transverse Cartesian coordinates x and y, and (iii) have phase fronts (surfaces of constant phase) that are spheres with radius of curvature R_z:

$$\psi_{mn}(x, y, z) = \frac{e^{-(x^2+y^2)/\sigma_z^2}}{\sqrt{2^{m+n-1}\pi\, m!n!}\,\sigma_z} H_m\left(\frac{\sqrt{2}\,x}{\sigma_z}\right) H_n\left(\frac{\sqrt{2}\,y}{\sigma_z}\right)$$

$$\times \exp\left\{i\left[\frac{k(x^2+y^2)}{2R_z} + kz - (n+m+1)\tan^{-1}\frac{z}{z_0}\right]\right\}. \quad (2)$$

Here $H_n(\xi) = (-1)^n e^{\xi^2} d^n e^{-\xi^2}/d\xi^n$ is the Hermite polynomial of index n, and m and n range over nonnegative integers. By adjusting σ_0, we can make the phase-front radius of curvature R_z match that, R, of the mirrors at the mirror locations, $z = \pm L/2$. Then the u_{mn} are a transversely orthonormal set of modes for the light field inside the cavity. Their flux distribution $|u_{mn}|^2$ on each mirror consists of an $m + 1$ by $n + 1$ matrix of discrete spots; see panel b in the box figure. The mode with $m = n = 0$ (panel a) is the Gaussian beam explored in the previous chapter: Eqs. (8.40). A given mode, specified by $\{m, n, k\}$, cannot resonate inside the cavity unless its wave number matches the cavity length, in the sense that the total phase shift in traveling from the cavity center $z = 0$ to the cavity end ($z = \pm L/2$) is an integral multiple of $\pi/2$; that is, $kL/2 - (n+m+1)\tan^{-1}[L/(2z_0)] = N\pi/2$ for some integer N.

There is a second family of modes that can resonate in the cavity, one whose eigenfunctions separate in circular polar coordinates:

$$\psi_{pm}(\varpi, \phi, z) = \frac{2p!e^{-\varpi^2/\sigma_z^2}}{\sqrt{1+\delta_{m0}}\,\pi(p+m)!\,\sigma_z}\left(\frac{\sqrt{2}\varpi}{\sigma_z}\right)^m L_p^m\left(\frac{2\varpi^2}{\sigma_z^2}\right)\begin{pmatrix}\cos m\phi \text{ or}\\ \sin m\phi\end{pmatrix}$$

$$\times \exp\left\{i\left[\frac{k\varpi^2}{2R_z} + kz - (2p+m+1)\tan^{-1}\frac{z}{z_0}\right]\right\}, \quad (3)$$

(continued)

BOX 9.3.   (continued)

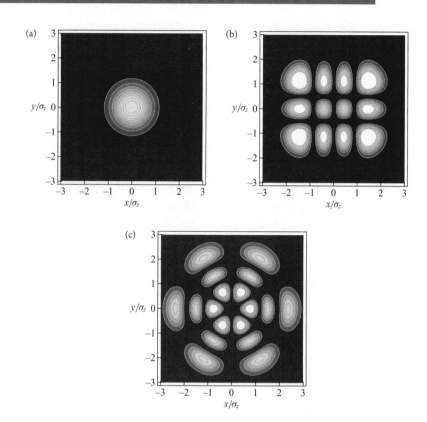

Energy flux distributions for (a) the Gaussian mode, (b) the Hermite mode of order 3,2, (c) the associated Laguerre mode of order 2,3. The contours are at 90%, 80%, . . . , 10% of maximum.

where $L_p^m(\xi) = (1/p!)\, e^\xi\, \xi^{-m}\, d^p/d\xi^p (e^{-\xi}\, \xi^{p+m})$ is the associated Laguerre polynomial, and the indices p and m range over nonnegative integers. These modes make spots on the mirrors shaped like azimuthal wedges cut radially by circles; see panel c in the box figure. Again, they can resonate only if the phase change from the center of the cavity to an end mirror at $z = \pm L/2$ is an integral multiple of $\pi/2$: $kL/2 - (2p + m + 1)\tan^{-1}[L/(2z_0)] = N\pi/2$.

As one goes to larger mode numbers m, n (Hermite modes) or p, m (associated Laguerre modes), the region with substantial light power gets larger (see the box figure). As a result, more light gets lost off the edges of the cavity's mirrors. So unless the mirrors are made very large, high-order modes have large losses and do not resonate well.

For further details on these modes, see, for example, Yariv and Yeh (2007, Secs. 2.5–2.8 and 4.3).

[Eqs. (9.33a) with $\theta = 0$ so the light rays are orthogonal to the mirrors]. Here $\mathfrak{r}$ is the mirrors' reflection coefficient, and the round-trip phase is now

$$\varphi = 2\pi\omega/\omega_f + \varphi_G, \quad \text{where } \omega_f = 2\pi/\tau_{\text{rt}}. \tag{9.37}$$

Here τ_{rt} is the time required for a high-frequency photon to travel round-trip in the interferometer along the optic axis, from one mirror to the other; ω_f (called the *free spectral range*) is, as we shall see, the angular-frequency separation between the interferometer's resonances; and φ_G is an additive contribution (the Gouy phase), caused by the curvature of the phase fronts [e.g., the $\tan^{-1}(z/z_o)$ term in Eq. (8.40a) for a Gaussian beam and in Eqs. (2) and (3) of Box 9.3 for higher-order modes]. Because φ_G is of order one while $2\pi\omega/\omega_f$ is huge compared to one, and because φ_G changes very slowly with changing light frequency, it is unimportant in principle (and we henceforth shall ignore it). However, it is important in practice: it causes modes with different transverse light distributions (e.g., the Gaussian and higher-order modes in Box 9.3), which have different Gouy phases, to resonate at different frequencies.

free spectral range

Gouy phase

The Fabry-Perot interferometer's power transmissivity T and reflectivity R are given by Eqs. (9.33c), which we can rewrite in the following simplified form:

$$T = 1 - R = \frac{1}{1 + (2\mathcal{F}/\pi)^2 \sin^2 \frac{1}{2}\varphi}. \tag{9.38}$$

Here $\mathcal{F}$, called the interferometer's *finesse*, is defined by

$$\mathcal{F} \equiv \pi\mathfrak{r}/(1 - \mathfrak{r}^2). \tag{9.39}$$

finesse

[This finesse should not be confused with the *coefficient of finesse* $F = (2\mathcal{F}/\pi)^2$, which is sometimes used in optics, but which we eschew to avoid confusion.]

In Fig. 9.9 we plot, as functions of the round-trip phase $\varphi = 2\pi\omega/\omega_f$ (ignoring φ_G), the interferometer's power reflectivity and transmissivity T and R, and the phase changes $\arg(\mathfrak{t}_{\text{FP}})$ and $\arg(\mathfrak{r}_{\text{FP}})$ of the light that is transmitted and reflected by the interferometer.

Notice in Fig. 9.9a that, when the finesse $\mathcal{F}$ is large compared to unity, the interferometer exhibits sharp resonances at frequencies separated by the free spectral range ω_f. On resonance, the interferometer is perfectly transmitting ($T = 1$); away from resonance, it is nearly perfectly reflecting ($R \simeq 1$). The *full width at half maximum* (FWHM) of each sharp transmission resonance is given by

$$\delta\varphi_{1/2} = \frac{2\pi}{\mathcal{F}}, \qquad \delta\omega_{1/2} = \frac{\omega_f}{\mathcal{F}}. \tag{9.40a}$$

resonance FWHM

In other words, if the frequency ω of the light is swept slowly through resonance, the transmission will be within 50% of its peak value (unity) over a bandwidth $\delta\omega_{1/2} = \omega_f/\mathcal{F}$. Notice also, in Fig. 9.9b, that for large finesse, the phase of the reflected and

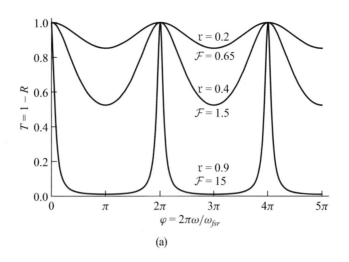

(a)

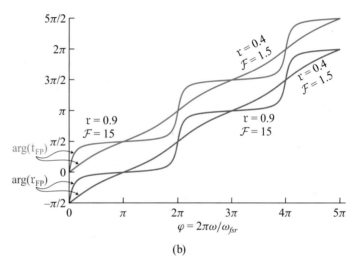

(b)

FIGURE 9.9 (a) Power transmissivity and reflectivity [Eq. (9.38)] for a Fabry-Perot interferometer with identical mirrors that have reflection coefficients $\mathfrak{r}$, as a function of the round-trip phase shift φ inside the interferometer. (b) The phase of the light transmitted (red) or reflected (blue) from the interferometer, relative to the input phase [Eqs. (9.36)]. The interferometer's finesse $\mathcal{F}$ is related to the mirrors' reflectivity by Eq. (9.39).

transmitted light near resonance changes very rapidly with a change in frequency of the driving light. Precisely on resonance, that rate of change is

resonance rate of change of phase

$$\left(\frac{d\,\arg(\mathfrak{t}_{FP})}{d\omega}\right)_{\text{on resonance}} = \left(\frac{d\,\arg(\mathfrak{r}_{FP})}{d\omega}\right)_{\text{on resonance}} = \frac{2\mathcal{F}}{\omega_f} = \frac{2}{\delta\omega_{1/2}}. \quad (9.40b)$$

The large transmissivity at resonance for large finesse can be understood by considering what happens when one first turns on the incident wave. Since the reflectivity of the first (input) mirror is near unity, the incoming wave has a large amplitude for reflection, and correspondingly, only a tiny amplitude for transmission into the optical cavity. The tiny bit that gets transmitted travels through the first mirror, gets strongly reflected from the second mirror, and returns to the first precisely in phase with the incoming wave (because φ is an integer multiple of 2π). Correspondingly, it superposes coherently on the tiny field being transmitted by the incoming wave, and so the net wave inside the cavity is doubled. After one more round trip inside the slab, this wave returns to the first face again in phase with the tiny field being transmitted by the incoming wave; again they superpose coherently, and the internal wave now has a three-times-larger amplitude than it began with. This process continues until a very strong field has built up inside the cavity (Ex. 9.12). As it builds up, that field begins to leak out of the cavity's first mirror with just such a phase as to destructively interfere with the wave being reflected there. The net reflected wave is thereby driven close to zero. The field leaking out of the second mirror has no other wave to interfere with. It remains strong, so the interferometer settles down into a steady state with strong net transmission. Heuristically, one can say that, because the wave inside the cavity is continually constructively superposing on itself, the cavity sucks almost all the incoming wave into itself, and then ejects it out the other side. Quantum mechanically, this sucking is due to the photons' Bose-Einstein statistics: the photons "want" to be in the same quantum state. We shall study this phenomenon in the context of plasmons that obey Bose-Einstein statistics in Sec. 23.3.2.

superposition of internal waves on resonance

Bose-Einstein behavior on resonance

This discussion makes it clear that, when the properties of the input light are changed, a high-finesse Fabry-Perot interferometer will change its response rather slowly—on a timescale approximately equal to the inverse of the resonance FWHM (i.e., the finesse times the round-trip travel time):

$$\tau_{\text{response}} \sim \frac{2\pi}{\delta\omega_{1/2}} = \mathcal{F}\,\tau_{\text{rt}}. \tag{9.40c}$$

These properties of a high-finesse Fabry-Perot interferometer are similar to those of a high-Q mechanical or electrical oscillator. The similarity arises because, in both cases, energy is being stored in a resonant, sinusoidal manner inside the device (the oscillator or the interferometer). For the interferometer, the light's round-trip travel time τ_{rt} is analogous to the oscillator's period, the interferometer's free spectral range ω_f is analogous to the oscillator's resonant angular frequency, and the interferometer's finesse $\mathcal{F}$ is analogous to the oscillator's quality factor Q. However, there are some major differences between an ordinary oscillator and a Fabry-Perot interferometer. Perhaps the most important is that the interferometer has several large families of resonant modes (families characterized by the number of longitudinal nodes between the mirrors and by the 2-dimensional transverse distributions of the light), whereas an oscillator has just one mode. This gives an interferometer much greater versatility than a simple oscillator possesses.

similarity between Fabry-Perot interferometer and an oscillator

9.4.3 Fabry-Perot Applications: Spectrometer, Laser, Mode-Cleaning Cavity, Beam-Shaping Cavity, PDH Laser Stabilization, Optical Frequency Comb T2

Just as mechanical and electrical oscillators have a wide variety of important applications in science and technology, so also do Fabry-Perot interferometers. In this section, we sketch a few of them.

SPECTROMETER

In the case of a Fabry-Perot etalon (highly reflecting parallel mirrors; plane-parallel light beam), the resonant transmission enables the etalon to be used as a spectrometer. The round-trip phase change $\varphi = 2n\omega d \cos\theta/c$ inside the etalon varies linearly with the wave's angular frequency ω, but only waves with round-trip phase φ near an integer multiple of 2π will be transmitted efficiently. The etalon can be tuned to a particular frequency by varying either the slab width d or the angle of incidence of the radiation (and thence the angle θ inside the etalon). Either way, impressively good chromatic resolving power can be achieved. We say that waves with two nearby frequencies can just be resolved by an etalon when the half-power point of the transmission coefficient of one wave coincides with the half-power point of the transmission coefficient of the other. Using Eq. (9.38) we find that the phases for the two frequencies must differ by $\delta\varphi \simeq 2\pi/\mathcal{F}$. Correspondingly, since $\varphi = 2n\omega d \cos\theta/c$, the *chromatic resolving power* is

chromatic resolving power of a spectrometer

$$\mathcal{R} = \frac{\omega}{\delta\omega} = \frac{4\pi n d \cos\theta}{\lambda_{\text{vac}} \delta\varphi} = \frac{2nd \cos\theta \, \mathcal{F}}{\lambda_{\text{vac}}}. \tag{9.41}$$

Here $\lambda_{\text{vac}} = 2\pi c/\omega$ is the wavelength in vacuum (i.e., outside the etalon).

LASER

Fabry-Perot interferometers are exploited in the construction of many types of lasers. For example, in a gas-phase laser, the atoms are excited to emit a spectral line. This radiation is spontaneously emitted isotropically over a wide range of frequencies. Placing the gas between the mirrors of a Fabry-Perot interferometer allows one or more highly collimated and narrowband modes to be trapped and, while trapped, to be amplified by stimulated emission (i.e., to lase). See Sec. 10.2.1.

MODE CLEANER FOR A MESSY LASER BEAM

The output beam from a laser often has a rather messy cross sectional profile, for example because it contains multiple modes of excitation of the Fabry-Perot interferometer in which the lasing material resides (see the discussion of possible modes in Box 9.3). For many applications, one needs a much cleaner laser beam (e.g., one with

cleaning a light beam

a Gaussian profile). To clean the beam, one can send it into a high-finesse Fabry-Perot cavity with identical spherical mirrors, whose mirror curvatures and cavity length are adjusted so that, among the modes present in the beam, only the desired Gaussian mode will resonate and thereby be transmitted (see the sharp transmission peaks in Fig. 9.9a). The beam's unwanted modes will not resonate in the cavity, and therefore will be reflected backward off its input mirror, leaving the transmitted beam clean.

BEAM-SHAPING CAVITY

In some applications one wants a light beam whose cross sectional distribution of flux $F(x, y)$ is different from any of the modes that resonate in a spherical-mirror cavity—for example, one might want a circular, flat-topped flux distribution $F(\varpi)$ with steeply dropping edges, like the shape of a circular mesa in a North American desert. One can achieve the desired light distribution (or something approximating it) as follows. Build a Fabry-Perot cavity with identical mirrors that are shaped in such a way that there is a cavity mode with the desired distribution. Then drive the cavity with a Gaussian beam. That portion of the beam with the desired flux distribution will resonate in the interferometer and leak out of the other mirror as the desired beam; the rest of the input beam will be rejected by the cavity.

reshaping a light beam

LASER STABILIZATION

There are two main ways to stabilize the frequency of a laser. One is to lock it onto the frequency of some fundamental atomic or molecular transition. The other is to lock it onto a resonant frequency of a mechanically stable Fabry-Perot cavity—a technique called *Pound-Drever-Hall* (PDH) *locking*.

PDH locking

In PDH locking to a cavity with identical mirrors, one passes the laser's output light (with frequency ω) through a device that modulates its frequency,[9] so ω becomes $\omega + \delta\omega$ with $\delta\omega = \sigma \cos(\Omega t)$ and $\sigma \ll \delta\omega_{1/2}$, the cavity resonance's FWHM. One then sends the modulated light into the cavity and monitors the reflected light power. Assume, for simplicity, that the modulation is slow compared to the cavity response time, $\Omega \ll 1/\tau_{\text{response}}$. Then the cavity's response at any moment will be that for steady light, that is, the reflected power will be $P_i R(\omega + \delta\omega)$, where R is the reflectivity at frequency $\omega + \delta\omega$. Using Eq. (9.38) for the reflectivity, specialized to frequencies very near resonance so the denominator is close to one, and using Eqs. (9.37) and (9.40a), we bring the reflected power into the form

$$P_r = P_i R(\omega + \delta\omega) = P_i \times \left[R(\omega) + \frac{dR}{d\omega} \delta\omega(t) \right]$$

$$= P_i \left[R(\omega) + \frac{8\sigma(\omega - \omega_o)}{(\delta\omega_{1/2})^2} \cos \Omega t \right], \qquad (9.42)$$

where ω_o is the cavity's resonant frequency (at which φ is an integral multiple of 2π).

The modulated part of the reflected power has an amplitude directly proportional to the laser's frequency error, $\omega - \omega_o$. In the PDH technique, one monitors this modulated power with a photodetector, followed by a band-pass filter on the photodetector's output electric current to get rid of the unmodulated part of the signal [arising from $P_i R(\omega)$]. The amplitude of the resulting, modulated output current is proportional

---

9. Actually, one sends it through a phase modulator called a *Pockels cell,* consisting of a crystal whose index of refraction is modulated by applying an oscillating voltage to it. The resulting phase modulation, $\delta\phi \propto \sin \Omega t$, is equivalent to a frequency modulation, $\delta\omega = d\delta\phi/dt \propto \cos \Omega t$.

to $\omega - \omega_o$ and is used to control a *feedback circuit* that drives the laser back toward the desired, cavity-resonant frequency ω_o. (See, e.g., Black, 2001, for details.)

In Ex. 9.15 it is shown that, if one needs a faster feedback and therefore requires a modulation frequency $\Omega \gtrsim 1/\tau_{\text{response}}$, this PDH locking technique still works.

This technique was invented by Ronald Drever for use in interferometric gravitational-wave detectors, relying on earlier ideas of Robert Pound, and it was first demonstrated experimentally by Drever and John Hall. It is now used widely in many areas of science and technology.

OPTICAL FREQUENCY COMB

John Hall and Theodor Hänsch were awarded the 2005 Nobel Prize for development of the optical frequency comb. This powerful tool is based on an optical cavity of length L, filled with a lasing medium that creates and maintains a sharply pulsed internal light field with the following form (Fig. 9.10):

$$\psi = \psi_o(z - V_g t) \exp[i k_p (z - V_p t)]. \tag{9.43}$$

Here

1. we use a z coordinate that increases monotonically along the optic axis as one moves rightward from mirror 1 to mirror 2, then leftward from 2 to 1, then rightward from 1 to 2, and so forth;

2. $\exp[i k_p(z - V_p t)]$ is the pth longitudinal monochromatic mode of the cavity, with wave number $k_p \equiv p\pi/L$, phase velocity V_p, and angular frequency $k_p V_p$ lying in the optical range $\sim 10^{15}$ Hz;

3. $\psi_o(z - V_g t)$ is the envelope of a wave packet so narrow that only about one wavelength of the mode p can fit inside it;

4. the envelope travels with group velocity V_g and does not spread.

For the wave packet not to spread, V_g must have the same (constant) value over all frequencies contained in the packet, which means that the dispersion relation must have $0 = (\partial V_g/\partial k) = \partial^2 \omega/\partial k^2$; ω must be linear in k:

$$\omega = V_g(k + \kappa) \tag{9.44}$$

for some constant κ, which is typically considerably smaller than the wave numbers k contained in the packet.

It was a huge technical challenge to build a lasing cavity that creates and sustains a very narrow wave packet of this sort. Two of the keys to this achievement were (i) using a nonlinear lasing medium that amplifies light more strongly at high energy fluxes $|\psi|^2$ than at low ones and thereby tends to produce intense, short pulses rather than long, monochromatic waves and (ii) using some trickery to ensure that the lasing medium and anything else in the cavity jointly give rise to the linear dispersion relation (9.44) over a sufficiently wide frequency band. For some of the details, see Sec. 10.2.3. Because the sharp wave packet (9.43) has fixed relationships between the phases of the

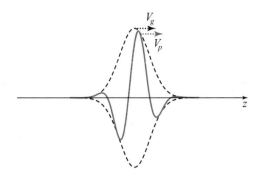

FIGURE 9.10 The sharply pulsed electric field [Eq. (9.43)] inside a Fabry-Perot cavity. The envelope is shown dotted; the red curve is the full field ψ.

various monochromatic modes that make it up, the lasing optical cavity that creates it is called a *mode-locked laser*.

mode-locked laser

As the internal field's pulse hits mirror 2 time and again, it transmits through the mirror a sequence of outgoing pulses separated by the wave packet's round-trip travel time in the cavity, $\tau_{rt} = 2L/V_g$. Assuming, for pedagogical clarity, a Gaussian shape for each pulse, the oscillating internal field (9.43) produces the outgoing field $\psi \propto \sum_n \exp[-\sigma^2(t - z/c - n\tau_{rt})^2/2] \exp[-ik_p V_p(t - z/c)]$. Here $1/\sigma$ is the pulse length in time, and we have assumed vacuum-light-speed propagation outside the cavity. It is helpful to rewrite the frequency $k_p V_p$ of the oscillatory piece of this field as the sum of its nearest multiple of the cavity's free spectral range, $\omega_f = 2\pi/\tau_{rt}$, plus a frequency shift ω_s: $k_p V_p = q\omega_f + \omega_s$. The integer $q =$ (largest integer contained in $k_p V_p/\omega_f$) will typically be quite close to p, and ω_s is guaranteed to lie in the interval $0 \le \omega_s < \omega_f$. The emerging electric field is then

$$\psi \propto \sum_n \exp[-\sigma^2(t - z/c - n\tau_{rt})^2/2] \exp[-i(q\omega_f + \omega_s)(t - z/c)] \quad \text{(9.45a)}$$

pulsed electric field from mode-locked laser

(Fig. 9.11a).

This entire emerging field is periodic in $t - z/c$ with period $\tau_{rt} = 2\pi/\omega_f$, except for the frequency-shift term $\exp[-i\omega_s(t - z/c)]$. The periodic piece can be expanded as a sum over discrete frequencies that are multiples of $\omega_f = 2\pi/\tau_{rt}$. Since the Fourier transform of a Gaussian is a Gaussian, this sum, augmented by the frequency shift term, turns out to be

$$\psi \propto \sum_{m=-\infty}^{+\infty} \exp\left[\frac{-(m - q)^2 \omega_f^2}{2\sigma^2}\right] \exp[-i(m\omega_f + \omega_s)(t - z/c)]. \quad \text{(9.45b)}$$

The set of discrete frequencies (spectral lines) appearing in this outgoing field is the *frequency comb* displayed in Fig. 9.11b.

electric field's optical frequency comb

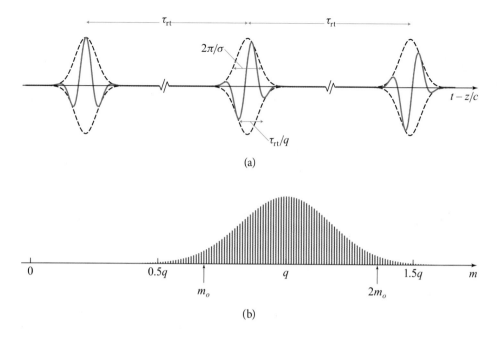

FIGURE 9.11 Optical frequency comb. (a) The pulsed electric field (9.45a) emerging from the cavity. (b) The field's comb spectrum. Each line, labeled by m, has angular frequency $m\omega_f$, and is shown with height proportional to its power, $\propto \exp\left[-(m-q)^2\omega_f^2/\sigma^2\right]$ [cf. Eq. (9.45b)].

Some concrete numbers make clear how very remarkable this pulsed electric field (9.45a) and its frequency comb (9.45b) are:

1. The Fabry-Perot cavity typically has a length L somewhere between ~ 3 cm and ~ 3 m, so (with the group velocity V_g of order the vacuum light speed) the round-trip travel time and free spectral range are $\tau_{rt} = 2L/V_g \sim 0.3$ to 30 ns; and $\omega_f/2\pi \sim 1/\tau_{rt} \sim 30$ MHz to 3 GHz, which are radio and microwave frequencies, respectively. Since the shift frequency is $\omega_s < \omega_f$, it is also in the radio or microwave band.

2. The comb's central frequency is in the optical, $q\omega_f/2\pi \sim 3 \times 10^{14}$ Hz, so the harmonic number of the central frequency is $q \sim 10^5$ to 10^7, roughly a million.

3. The pulse width $\sim 2/\sigma$ contains roughly one period $(2\pi/q\omega_f)$ of the central frequency, so $\sigma \sim q\omega_f/3$, which means that most of the comb's power is contained in the range $m \sim 2q/3$ to $m \sim 4q/3$ (i.e., there are roughly a million strong teeth in the comb).

It is possible to lock the comb's free spectral range ω_f to a very good cesium atomic clock, whose oscillation frequency ~ 9 GHz is stable to $\delta\omega/\omega \sim 10^{-13}$ (Fig. 6.11), so ω_f has that same phenomenal stability. One can then measure the shift frequency ω_s

and calibrate the comb (identify the precise mode number m of each frequency in the comb) as follows.

1. Arbitrarily choose a tooth at the low-frequency end of the comb, $m_o \simeq 2q/3$ (Fig. 9.11b); it has frequency $\omega_o = m_o \omega_f + \omega_s$.

2. Separate the light in that tooth from light in the other teeth, and send a beam of that tooth's light through a frequency doubler (to be discussed in Sec. 10.7.1), thereby getting a beam with frequency $2\omega_o = 2(m_o \omega_f + \omega_s)$.

3. By beating this beam against the light in teeth at $m \sim 4q/3$, identify the tooth that most closely matches this beam's frequency. It will have frequency $2m_o \omega_f + \omega_s$, and the frequency difference (beat frequency) will be ω_s.

This reveals ω_s to very high accuracy, and one can count the number of teeth $(m_o - 1)$ between this tooth $2m_o$ and its undoubled parent m_o, thereby learning the precise numerical value of m_o. From this, by tooth counting, one learns the precise mode numbers m of all the optical-band teeth in the comb, and also their frequencies $m\omega_f + \omega_s$.

With the comb now calibrated, it can be used to measure the frequency of any other beam of light in the optical band in terms of the ticking frequency of the cesium clock, to which the entire comb has been locked. The optical frequency accuracies thereby achieved are orders of magnitude better than were possible before this optical frequency comb was developed. And in the near future, as optical-frequency atomic clocks become much more accurate and stable than the microwave-frequency cesium clock (see the footnote in Sec. 6.6.1), this comb will be used to calibrate microwave and radio frequencies in terms of the ticking rates of optical frequency clocks.

For further details about optical frequency combs, see the review articles by Cundiff (2002) and Cundiff and Ye (2003).

EXERCISES

Exercise 9.15 *Problem: PDH Laser Stabilization* T2
Show that the PDH method for locking a laser's frequency to an optical cavity works for modulations faster than the cavity's response time, $\Omega \gtrsim 1/\tau_{\text{response}}$, and even works for $\Omega \gg 1/\tau_{\text{response}}$. More specifically, show that the reflected power still contains the information needed for feedback to the laser. [For a quite general analysis and some experimental details, see Black (2001).]

Exercise 9.16 *Derivation: Optical Frequency Comb* T2
Fill in the details of the derivation of all the equations in the section describing the optical frequency comb.

Exercise 9.17 ***Problem: Sagnac Interferometer* T2
A Sagnac interferometer is a rudimentary version of a laser gyroscope for measuring rotation with respect to an inertial frame. The optical configuration is shown in

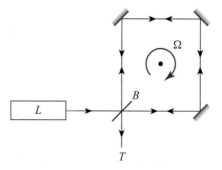

FIGURE 9.12 Sagnac interferometer used as a type of laser gyro.

Fig. 9.12. Light from a laser L is split by a beam splitter B and travels both clockwise and counterclockwise around the optical circuit, reflecting off three plane mirrors. The light is then recombined at B, and interference fringes are viewed through the telescope T. The whole assembly rotates with angular velocity Ω. Calculate the difference in the time it takes light to traverse the circuit in the two directions, and show that the consequent fringe shift (total number of fringes that enter the telescope during one round trip of the light in the interferometer) can be expressed as $\Delta N = 4A\Omega/(c\lambda)$, where λ is the wavelength, and A is the area bounded by the beams. Show further that, for a square Sagnac interferometer with side length L, the rate at which fringes enter the telescope is $\Omega L/\lambda$.

9.5   9.5 Laser Interferometer Gravitational-Wave Detectors T2

As we discuss in Chap. 27, gravitational waves are predicted to exist by general relativity theory, and their emission by the binary neutron-star system PSR B1913+16 has been monitored since 1974, via their back-action on the binary's orbital motion. As orbital energy is lost to gravitational waves, the binary gradually spirals inward, so its orbital angular velocity gradually increases. The measured rate of increase agrees with general relativity's predictions to within the experimental accuracy of a fraction of a percent (for which Russell Hulse and Joseph Taylor received the 1993 Nobel Prize in physics). Unfortunately, the gravitational analog of Hertz's famous laboratory emission and detection of electromagnetic waves has not yet been performed, and cannot be in the authors' lifetime because of the waves' extreme weakness. For waves strong enough to be detectable on Earth, one must turn to violent astrophysical events, such as the collision and coalescence of two neutron stars or black holes.

    When the gravitational waves from such an event reach Earth and pass through a laboratory, general relativity predicts that they will produce tiny relative accelerations of free test masses. The resulting oscillatory variation of the distance between two such masses can be measured optically using a Michelson interferometer, in which (to increase the signal strength) each of the two arms is operated as a Fabry-Perot

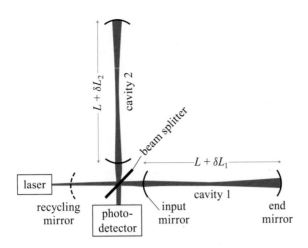

FIGURE 9.13 Schematic design of an initial gravitational-wave interferometer operated in LIGO (at Livingston, Louisiana, and Hanford, Washington, USA) during 2005–2010.

cavity. Two such instruments, called *interferometric gravitational wave detectors* [for which Rainer Weiss (1972) was the primary inventor, with important contributions from Ronald Drever and many others] made the first discovery of gravitational waves arriving at Earth on September 14, 2015 (Abbott et al., 2016a). The waves came from the last few orbits of inspiral and the collision of two black holes with masses $29M_\odot$ and $36M_\odot$, where $M_\odot$ is the Sun's mass.

interferometric gravitational wave detector

The observatory that discovered these waves is called LIGO, the Laser Interferometer Gravitational Wave Observatory, and its interferometric detectors were second generation instruments, called advanced LIGO. These advanced LIGO detectors are more complex than is appropriate to analyze in this chapter, so we shall forcus instead on the first generation detectors called initial LIGO (Abbott et al., 2009), and then near the end of this section we shall briefly describe how advanced LIGO differs from initial LIGO.

advanced LIGO

initial LIGO

In each of the initial LIGO gravitational-wave detectors, the two cavities are aligned along perpendicular directions, as shown in Fig. 9.13. A Gaussian beam of light (Sec. 8.5.5) from a laser passes through a beam splitter, creating two beams with correlated phases. The beams excite the two cavities near resonance. Each cavity has an end mirror with extremely high reflectivity,[10] $1 - \mathfrak{r}_e^2 < 10^{-4}$, and a corner mirror ("input mirror") with a lower reflectivity, $1 - \mathfrak{r}_i^2 \sim 0.03$. Because of this lower reflectivity, by contrast with the etalons discussed in previous sections, the resonant light leaks out through the input mirror instead of through the end mirror. This reflectivity of the input mirror is so adjusted that the typical photon is stored in

mirror reflectivity

10. Because LIGO operates with monochromatic light, it is convenient to adjust the phases of the mirrors' reflection and transmission coefficients so $\mathfrak{r}$ and $\mathfrak{t}$ are both real; cf. Ex. 9.10e. We do so here.

the cavity for roughly half the period of the expected gravitational waves (a few milliseconds), which means that the input mirror's reflectivity $\mathfrak{r}_i^2$, the arm length L, and the gravitational-wave angular frequency ω_{gw} are related by

$$\frac{4L}{c(1 - \mathfrak{r}_i^2)} \sim \frac{1}{\omega_{gw}}. \tag{9.46}$$

The light emerging from the cavity, like that transmitted by an etalon, has a phase that is highly sensitive to the separation between the mirrors: a tiny change δL in their separation produces a change in the outcoming phase

$$\delta\varphi_o \simeq \frac{8\omega\delta L}{c} \frac{1}{(1 - \mathfrak{r}_i^2)} \sim \frac{2\omega}{\omega_{gw}} \frac{\delta L}{L} \tag{9.47}$$

in the limit $1 - \mathfrak{r}_i \ll 1$; see Ex. 9.18. The outcoming light beams from the two cavities return to the beam splitter and are recombined there. The relative distances from the beam splitter to the cavities are adjusted so that, in the absence of any perturbations of the cavity lengths, almost all the interfered light goes back toward the laser, and only a tiny (but nonzero) amount goes toward the photodetector of Fig. 9.13, which monitors the output. Perturbations δL_1 and δL_2 in the cavity lengths then produce a change

phase shift induced by mirror motions

$$\delta\varphi_{o1} - \delta\varphi_{o2} \sim \frac{2\omega}{\omega_{gw}} \frac{(\delta L_1 - \delta L_2)}{L} \tag{9.48}$$

in the relative phases at the beam splitter, and this in turn produces a change of the light power entering the photodetector. By using two cavities in this way and keeping their light-storage times (and hence response times) the same, one makes the light power entering the photodetector be insensitive to fluctuations in the laser frequency. This is crucial for obtaining the high sensitivities that gravitational-wave detection requires.

The mirrors at the ends of each cavity are suspended as pendula, and when a gravitational wave with dimensionless amplitude h (discussed in Chap. 27) passes, it moves the mirrors back and forth, producing changes

mirror motions produced by gravitational wave

$$\delta L_1 - \delta L_2 = hL \tag{9.49}$$

in the arm-length difference. The resulting change in the relative phases of the two beams returning to the beam splitter,

$$\delta\varphi_{o1} - \delta\varphi_{o2} \sim \frac{2\omega}{\omega_{gw}} h, \tag{9.50}$$

is monitored via the changes in power that it produces for the light going into the photodetector. If one builds the entire detector optimally and uses the best possible

detectable phase shift

photodetector, these phase changes can be measured with a photon shot-noise-limited precision of $\sim 1/\sqrt{N}$. Here $N \sim [W_\ell/(\hbar\omega)](\pi/\omega_{gw})$ is the number of photons put into the detector by the laser (with power W_ℓ) during half a gravitational-wave period.[11] By combining this with Eq. (9.50), we see that the weakest wave that can be detected (at signal-to-noise ratio 1) is

$$h \sim \left(\frac{\hbar \omega_{gw}^3}{4\pi \omega W_\ell} \right)^{1/2}. \qquad (9.51)$$

estimate of weakest detectable gravitational wave

For a laser power $W_\ell \sim 5$ W, and $\omega_{gw} \sim 10^3$ s^{-1}, $\omega \sim 2 \times 10^{15}$ s^{-1}, this gravitational-wave sensitivity (noise level) is $h \sim 1 \times 10^{-21}$.

When operated in this manner, about 97% of the light returns toward the laser from the beam splitter and the other 3% goes out the end mirror, goes into the photodetector, gets absorbed, or is scattered due to imperfections in the optics. In LIGO's initial detectors, the 97% returning toward the laser was recycled back into the interferometer, in phase with new laser light, by placing a mirror between the laser and the beam splitter. This "power recycling mirror" (shown dashed in Fig. 9.13) made the entire optical system into a big optical resonator with two subresonators (the arms' Fabry-Perot cavities), and the practical result was a 50-fold increase in the input light power, from 5 W to 250 W—and an optical power in each arm of about $\frac{1}{2} \times 250$ W $\times 4/(1 - \mathfrak{r}_i^2) \sim 15$ kW; see Abbott et al. (2009, Fig. 3). When operated in this manner, the interferometer achieved a noise level $h \sim 1 \times 10^{-21}/\sqrt{50} \sim 1 \times 10^{-22}$. For a more accurate analysis of the sensitivity, see Exs. 9.18 and 9.19.

power recycling

This estimate of sensitivity is actually the rms noise in a bandwidth equal to frequency at the minimum of LIGO's noise curve. Figure 6.7 shows the noise curve as the square root of the spectral density of the measured arm-length difference $\xi \equiv L_1 - L_2$, $\sqrt{S_\xi(f)}$. Since the waves produce a change of ξ given by $\delta\xi = hL$ [Eq. (9.49)], the corresponding noise-induced fluctuations in the measured h have $S_h = S_\xi/L^2$, and the rms noise fluctuations in a bandwidth equal to frequency f are $h_{rms} = \sqrt{S_h f} = (1/L)\sqrt{S_\xi f}$. Inserting $\sqrt{S_\xi} \simeq 10^{-19}$ m Hz$^{-1/2}$ and $f \simeq 100$ Hz from Fig. 6.7, and $L = 4$ km for the LIGO arm length, we obtain $h_{rms} \simeq 2.5 \times 10^{-22}$, a factor 2.5 larger than our 1×10^{-22} estimate—in part because thermal noise is roughly as large as photon shot noise at 100 Hz.

measured noise in initial LIGO

There are enormous obstacles to achieving such high sensitivity. Here we name just a few. Imperfections in the optics will absorb some of the high light power, heating the mirrors and beam splitter and causing them to deform. Even without such heating, the mirrors and beam splitter must be exceedingly smooth and near perfectly shaped to minimize the scattering of light from them (which causes noise; Ex. 8.15). Thermal noise in the mirrors and their suspensions (described by the fluctuation-dissipation

experimental challenges

---

11. This measurement accuracy is related to the Poisson distribution of the photons entering the interferometer's two arms: if N is the mean number of photons during a half gravitational-wave period, then the variance is $\sqrt{N}$, and the fractional fluctuation is $1/\sqrt{N}$. The interferometer's shot noise is actually caused by a beating of quantum electrodynamical vacuum fluctuations against the laser's light; for details see Caves (1980).

theorem) will cause the mirrors to move in manners that simulate the effects of a gravitational wave (Secs. 6.8.2 and 11.9.2), as will seismic- and acoustic-induced vibrations of the mirror suspensions. LIGO's arms must be long (4 km) to minimize the effects of these noise sources. While photon shot noise dominates above the noise curve's minimum, $f \gtrsim 100$ Hz, these and other noises dominate at lower frequencies.

The initial LIGO detectors operated from 2005 to 2010, carrying out gravitational-wave searches, much of the time in collaboration with international partners (the French-Italian VIRGO and British/German GEO600 detectors). From 2010 to 2015, the second generation advanced LIGO detectors were installed in the LIGO vacuum system, with amplitude design sensitivity 10-fold higher than the initial detectors. These advanced detectors discovered gravitational waves in September 2015, just before their first search officially began, and when their sensitivity at the noise-curve minimum was about 3 times worse than their design but 3 times better than initial LIGO.

advanced LIGO noise and discovery of gravitational waves

One of several major changes, in going from initial LIGO to advanced LIGO, was the insertion of a "signal recycling mirror" between the beam splitter and the photodetector (Abbott et al., 2016b). The output signal light, in the cavity bounded by this new mirror and the two input mirrors, "sucks" signal light out of the interferometer arms due to Bose-Einstein statistics. This permits laser light to be stored in the arm cavities far longer than half a gravitational wave period (and thereby build up to higher intensity), while the signal light gets extracted in roughly a half period. This added complication is why we chose not to analyze the advanced detectors.

advanced LIGO's signal recycling mirror

EXERCISES

Exercise 9.18 *Derivation and Problem: Phase Shift in LIGO Arm Cavity* T2

In this exercise and the next, simplify the analysis by treating each Gaussian light beam as though it were a plane wave. The answers for the phase shifts will be the same as for a true Gaussian beam, because on the optic axis, the Gaussian beam's phase [Eq. (8.40a) with $\varpi = 0$] is the same as that of a plane wave, except for the Gouy phase $\tan^{-1}(z/z_0)$, which is very slowly changing and thus is irrelevant.

(a) For the interferometric gravitational-wave detector depicted in Fig. 9.13 (with the arms' input mirrors having amplitude reflectivities $\mathfrak{r}_i$ close to unity and the end mirrors idealized as perfectly reflecting), analyze the light propagation in cavity 1 by the same techniques as used for an etalon in Sec. 9.4.1. Show that, if ψ_{i1} is the light field impinging on the input mirror, then the total reflected light field ψ_{r1} is

$$\psi_{r1} = e^{i\varphi_1} \frac{1 - \mathfrak{r}_i e^{-i\varphi_1}}{1 - \mathfrak{r}_i e^{i\varphi_1}} \psi_{i1}, \quad \text{where } \varphi_1 = 2kL_1. \tag{9.52a}$$

(b) From this, infer that the reflected flux $|\psi_{r1}|^2$ is identical to the cavity's input flux $|\psi_{i1}|^2$, as it must be, since no light can emerge through the perfectly reflecting end mirror.

(c) The arm cavity is operated on resonance, so φ_1 is an integer multiple of 2π. From Eq. (9.52a) infer that (up to fractional errors of order $1 - \mathfrak{r}_i$) a change δL_1 in the length of cavity 1 produces a change

$$\delta\varphi_{r1} = \frac{8k\delta L_1}{1 - \mathfrak{r}_i^2}. \tag{9.52b}$$

With slightly different notation, this is Eq. (9.47), which we used in the text's order-of-magnitude analysis of LIGO's sensitivity. In this exercise and the next, we carry out a more precise analysis.

Exercise 9.19 *Example: Photon Shot Noise in LIGO* T2

This exercise continues the preceding one. We continue to treat the light beams as plane waves.

(a) Denote by ψ_ℓ the light field from the laser that impinges on the beam splitter and gets split in two, with half going into each arm (Fig. 9.13). Using the equations from Ex. 9.18 and Sec. 9.5, infer that the light field returning to the beam splitter from arm 1 is $\psi_{s1} = \frac{1}{\sqrt{2}}\psi_\ell e^{i\varphi_1}(1 + i\delta\varphi_{r1})$, where φ_1 is some net accumulated phase that depends on the separation between the beam splitter and the input mirror of arm 1.

(b) Using the same formula for the field ψ_{s2} from arm 2, and assuming that the phase changes between beam splitter and input mirror are almost the same in the two arms, so $\varphi_o \equiv \varphi_1 - \varphi_2$ is small compared to unity (mod 2π), show that the light field that emerges from the beam splitter, traveling toward the photodetector, is

$$\psi_{pd} = \frac{1}{\sqrt{2}}(\psi_{s1} - \psi_{s2}) = \frac{i}{2}(\varphi_o + \delta\varphi_{r1} - \delta\varphi_{r2})\psi_\ell \tag{9.53a}$$

to first order in the small phases. Show that the condition $|\varphi_o| \ll 1$ corresponds to the experimenters' having adjusted the positions of the input mirrors in such a way that almost all the light returns toward the laser, and only a small fraction goes toward the photodetector.

(c) For simplicity, let the gravitational wave travel through the interferometer from directly overhead and have an optimally oriented polarization. Then, as we shall see in Chap. 27 [Eq. (27.81)], the dimensionless gravitational-wave field $h(t)$ produces the arm-length changes $\delta L_1 = -\delta L_2 = \frac{1}{2}h(t)L$, where L is the unperturbed arm length. Show, then, that the field traveling toward the photodetector is

$$\psi_{pd} = \frac{i}{2}(\varphi_o + \delta\varphi_{gw})\psi_\ell, \quad \text{where} \quad \delta\varphi_{gw} = \frac{8kL}{1 - \mathfrak{r}_i^2}h(t) = \frac{16\pi L/\lambda}{1 - \mathfrak{r}_i^2}h(t). \tag{9.53b}$$

The experimenter adjusts φ_o so it is large compared to the tiny $\delta\varphi_{gw}$.

(d) Actually, this equation has been derived assuming, when analyzing the arm cavities [Eq. (9.52a)], that the arm lengths are static. Explain why it should still be nearly valid when the gravitational waves are moving the mirrors, so long as the gravitational-wave half-period $1/(2f) = \pi/\omega_{gw}$ is somewhat longer than the mean time that a photon is stored inside an arm cavity, or so long as $f \ll f_o$, where

$$\boxed{f_o \equiv \frac{1-\mathfrak{r}_i^2}{4\pi}\frac{c}{2L}.} \tag{9.54}$$

Assume that this is so. For the initial LIGO detectors, $1 - \mathfrak{r}_i^2 \sim 0.03$ and $L = 4$ km, so $f_o \sim 90$ Hz.

(e) Show that, if W_ℓ is the laser power impinging on the beam splitter (proportional to $|\psi_\ell|^2$), then the steady-state light power going toward the photodetector is $W_{pd} = (\varphi_o/2)^2 W_\ell$, and the time variation in that light power due to the gravitational wave (the gravitational-wave signal) is

$$W_{gw}(t) = \sqrt{W_\ell W_{pd}}\,\frac{16\pi L/\lambda}{1-\mathfrak{r}_i^2}h(t). \tag{9.55a}$$

The photodetector monitors these changes $W_{gw}(t)$ in the light power W_{pd} and from them infers the gravitational-wave field $h(t)$. This is called a "DC" or "homodyne" readout system; it works by beating the gravitational-wave signal field ($\propto \delta\varphi_{gw}$) against the steady light field ("local oscillator," $\propto \varphi_o$) to produce the signal light power $W_{gw}(t) \propto h(t)$.

(f) Shot noise in the interferometer's output light power W_{pd} gives rise to noise in the measured gravitational-wave field $h(t)$. From Eq. (9.55a) show that the spectral density of the noise in the measured $h(t)$ is

$$S_h(f) = \left(\frac{(1-\mathfrak{r}_i^2)\lambda}{16\pi L}\right)^2\frac{S_{W_{pd}}}{W_\ell W_{pd}}. \tag{9.55b}$$

In Sec. 6.7.4, we derived the formula $S_{W_{pd}} = 2\mathcal{R}(\hbar\omega)^2 = 2W_{pd}\hbar\omega$ [Eq. (6.69)] for the (frequency-independent) spectral density of a steady, monochromatic light beam's power fluctuations due to shot noise; here $\mathcal{R} = W_{pd}/(\hbar\omega)$ is the average rate of arrival of photons. Combining with Eq. (9.55b), deduce your final formula for the spectral density of the noise in the inferred gravitational-wave signal:

$$S_h(f) = \left(\frac{(1-\mathfrak{r}_i^2)\lambda}{16\pi L}\right)^2\frac{2}{W_\ell/(\hbar\omega)}. \tag{9.56a}$$

From this deduce the rms noise in a bandwidth equal to frequency

$$\boxed{h_{rms} = \sqrt{f S_h} = \left(\frac{(1-\mathfrak{r}_i^2)\lambda}{16\pi L\sqrt{N}}\right),} \quad \text{where} \quad \boxed{N = \frac{W_\ell}{\hbar\omega}\frac{1}{2f}} \tag{9.56b}$$

is the number of photons that impinge on the beam splitter, from the laser, in half a gravitational-wave period.

(g) In Ex. 9.20 we shall derive (as a challenge) the modification to the spectral density that arises at frequencies $f \gtrsim f_o$. The signal strength that gets through the interferometer is reduced because the arm length is increasing, then decreasing, then increasing again (and so forth) while the typical photon is in an arm cavity. The result of the analysis is an increase of $S_h(f)$ by $1 + (f/f_o)^2$, so we have

$$S_h(f) = \left(\frac{(1 - \mathfrak{r}_i^2)\lambda}{16\pi L} \right)^2 \frac{2}{W_\ell/\hbar\omega} \left(1 + \frac{f^2}{f_o^2} \right).$$

(9.57)

Compare this with the measured noise, at frequencies above $f_o \sim 90$ Hz in the initial LIGO detectors (Fig. 6.7 with $\xi = hL$), using the initial LIGO parameters: $\lambda = 1.06 \ \mu$m, $\omega = 2\pi c/\lambda \simeq 2 \times 10^{15}$ s^{-1}, $L = 4$ km, $W_\ell = 250$ W, $1 - \mathfrak{r}_i^2 = 0.03$. It should agree fairly well with the measured noise at frequencies $f \gtrsim f_o$, where most of the noise is due to photon shot noise. Also compare the noise (9.57) in a bandwidth equal to frequency $\sqrt{f S_h}$ and evaluated at frequency $f = f_o$ with the estimate (9.51) worked out in the text.

Exercise 9.20 *Challenge: LIGO Shot Noise at $f \gtrsim f_o$* **T2**
Derive the factor $1 + (f/f_o)^2$ by which the spectral density of the shot noise is increased at frequencies $f \gtrsim f_o$. [Hint: Redo the analysis of the arm cavity fields, part (a) of Ex. 9.19, using an arm length that varies sinusoidally at frequency f due to a sinusoidal gravitational wave, and then use the techniques of Ex. 9.19 to deduce $S_h(f)$.]

9.6 Power Correlations and Photon Statistics:
Hanbury Brown and Twiss Intensity Interferometer

9.6

A type of interferometer that is rather different from those studied earlier in this chapter was proposed and constructed by Robert Hanbury Brown and Richard Q. Twiss in 1956. In this interferometer, light powers, rather than amplitudes, are combined to measure the degree of coherence of the radiation field. This is often called an *intensity interferometer*, because the optics community often uses the word "intensity" to mean energy flux (power per unit area).

intensity interferometer

In their original experiment, Hanbury Brown and Twiss divided light from an incandescent mercury lamp and sent it along two paths of variable length before detecting photons in each beam separately using a photodetector (see Fig. 9.14). The electrical output from each photodetector measures the rate of arrival of its beam's photons, or equivalently, its beam's power $W(t)$, which we can write as $K(\mathfrak{R}\Psi)^2$,

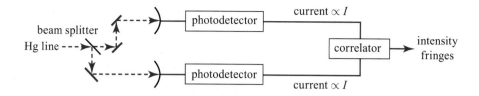

FIGURE 9.14 Hanbury Brown and Twiss intensity interferometer.

fluctuations of power W in each beam

where K is a constant. This W exhibits fluctuations δW about its mean value $\overline{W}$, and it was found that the fluctuations in the two beams were correlated. How can this be?

The light that was detected originated from many random and independent emitters and therefore obeys Gaussian statistics, according to the central limit theorem (Sec. 6.3.2). This turns out to mean that the fourth-order correlations of the wave field Ψ with itself can be expressed in terms of the second-order correlations (i.e., in terms of the degree of coherence $\gamma_{\parallel}$).

More specifically, continuing to treat the wave field Ψ as a scalar, we (i) write each beam's power as the sum over a set of Fourier components Ψ_j with precise frequencies ω_j and slowly wandering, complex amplitudes $W(t) = (\sum_j \Re\Psi_j)^2$; (ii) form the product $W(t)W(t+\tau)$; (iii) keep only terms that will have nonzero averages by virtue of containing products of the form $\propto e^{+i\omega_j t}e^{-i\omega_j t}e^{+i\omega_k t}e^{-i\omega_k t}$ (where j and k are generally not the same); and (iv) average over time. Thereby we obtain

$$\overline{W(t)W(t+\tau)} = K^2\overline{\Psi(t)\Psi^*(t)} \times \overline{\Psi(t+\tau)\Psi^*(t+\tau)}$$
$$+ K^2\overline{\Psi(t)\Psi^*(t+\tau)} \times \overline{\Psi^*(t)\Psi(t+\tau)}$$
$$= \overline{W}^2[1 + |\gamma_{\parallel}(\tau)|^2] \tag{9.58}$$

[cf. Eq. (9.16) with Ψ allowed to be complex]. If we now measure the relative fluctuations, we find that

longitudinal correlation of beam power

$$\frac{\overline{\delta W(t)\delta W(t+\tau)}}{\overline{W(t)}^2} = \frac{\overline{W(t)W(t+\tau)} - \overline{W(t)}^2}{\overline{W(t)}^2} = |\gamma_{\parallel}(\tau)|^2. \tag{9.59}$$

[Note: This analysis is only correct if the radiation comes from many uncorrelated sources—the many independently emitting mercury atoms in Fig. 9.14—and therefore has Gaussian statistics.]

Equation (9.59) tells us that the power as well as the amplitude of coherent radiation must exhibit a positive longitudinal correlation, and the degree of coherence for the fluxes is equal to the squared modulus of the degree of coherence for the amplitudes. Although this result was rather controversial at the time the experiments were first performed, it is easy to interpret qualitatively if we think in terms of photons rather than classical waves. Photons are bosons and are therefore positively correlated even in thermal equilibrium; cf. Chaps. 3 and 4. When they arrive at the beam splitter of Fig. 9.14, they clump more than would be expected for a random distribution of classical particles, a phenomenon called photon bunching.[12] In fact, treating the problem from the point of view of photon statistics gives an answer equivalent to Eq. (9.59).

explanation of power correlation by Bose-Einstein statistics: photon bunching

Some practical considerations should be mentioned. The first is that Eq. (9.59), derived for a scalar wave, is really only valid for electromagnetic waves if they are completely polarized. If the incident waves are unpolarized, then the intensity fluctuations are reduced by a factor of two. The second point is that, in the Hanbury Brown and Twiss experiments, the photon counts were actually averaged over longer times than the correlation time of the incident radiation. This reduced the magnitude of the measured effect further.

Nevertheless, after successfully measuring temporal power correlations, Hanbury Brown and Twiss constructed a *stellar intensity interferometer,* with which they were able to measure the angular diameters of bright stars. This method had the advantage that it did not depend on the phase of the incident radiation, so the results were insensitive to atmospheric fluctuations (seeing), one of the drawbacks of the Michelson stellar interferometer (Sec. 9.2.5). Indeed, it is not even necessary to use accurately ground mirrors to measure the effect. The method has the disadvantage that it can only measure the modulus of the degree of coherence; the phase is lost. It was the first example of using fourth-order correlations of the light field to extract image information from light that has passed through Earth's turbulent atmosphere (Box 9.2).

stellar intensity interferometer

Exercise 9.21 *Derivation: Power Correlations*
By expressing the field as either a Fourier sum or a Fourier integral complete the argument that leads to Eq. (9.58).

Exercise 9.22 *Problem: Electron Intensity Interferometry*
Is it possible to construct an intensity interferometer (i.e., a number-flux interferometer) to measure the coherence properties of a beam of electrons? What qualitative differences do you expect there to be from a photon-intensity interferometer? What do you expect Eq. (9.59) to become?

---

12. Pions emerging from high-energy, heavy ion collisions exhibit similar bunching because pions, like photons, are bosons.

Bibliographic Note

For pedagogical introductions to interference and coherence at an elementary level in greater detail than this chapter, see Klein and Furtak (1986) and Hecht (2017). For more advanced treatments, we like Pedrotti, Pedrotti, and Pedrotti (2007), Saleh and Teich (2007), Ghatak (2010), and especially Brooker (2003). For a particularly deep and thorough discussion of coherence, see Goodman (1985). For modern applications of interferometry (including the optical frequency comb), see Yariv and Yeh (2007), and at a more elementary level, Francon and Willmans (1966) and Hariharan (2007).

10

Nonlinear Optics

The development of the maser and laser . . . followed no script except to hew to
the nature of humans groping to understand, to explore, and to create . . .

CHARLES H. TOWNES (2002)

10.1 Overview

Communication technology is undergoing a revolution, and computer technology
may do so soon—a revolution in which the key devices used (e.g., switches and communication lines) are changing from radio and microwave frequencies to optical frequencies. This revolution has been made possible by the invention and development
of lasers (especially semiconductor diode lasers) and other technology developments,
such as nonlinear media whose polarization P_i is a nonlinear function of the applied
electric field, $P_i = \epsilon_0(\chi_{ij}E^j + 2d_{ijk}E^jE^k + 4\chi_{ijkl}E^jE^kE^l + \cdots)$. In this chapter we
study lasers, nonlinear media, and various nonlinear optics applications that are based
on them.

Most courses in elementary physics idealize the world as linear. From the simple
harmonic oscillator to Maxwell's equations to the Schrödinger equation, most elementary physical laws one studies are linear, and most of the applications studied make
use of this linearity. In the real world, however, nonlinearities abound, creating such
phenomena as avalanches, breaking ocean waves, holograms, optical switches, and
neural networks; and in the past three decades nonlinearities and their applications
have become major themes in physics research, both basic and applied. This chapter,
with its exploration of nonlinear effects in optics, serves as a first introduction to some
fundamental nonlinear phenomena and their present and future applications. In later
chapters, we revisit some of these phenomena and meet others, in the context of fluids
(Chaps. 16 and 17), plasmas (Chap. 23), and spacetime curvature (Chaps. 25–28).

Since highly coherent and monochromatic laser light is one of the key foundations
on which modern nonlinear optics has been built, we begin in Sec. 10.2 with a review
of the basic physics principles that underlie the laser: the pumping of an active medium
to produce a molecular population inversion and the stimulated emission of radiation
from the inverted population. Then we briefly describe the wide variety of lasers now
available, how a few of them are pumped, and the characteristics of their light. As

an important example (crucial for the optical frequency combs of Sec. 9.4.3), we give details about mode-locked lasers.

In Sec. 10.3, we meet our first example of an application of nonlinear optics: holography. In the simplest variant of holography, a 3-dimensional, monochromatic image of an object is produced by a two-step process: recording a hologram of the image, and then passing coherent light through the hologram to reconstruct the image. We analyze this recording and reconstruction and then describe a few of the many variants of holography now available and some of their practical applications.

Holography differs from more modern nonlinear optics applications in not being a real-time process. Real-time processes have been made possible by nonlinear media and other new technologies. In Sec. 10.4, we study an example of a real-time, nonlinear-optics process: phase conjugation of light by a phase-conjugating mirror (though we delay a detailed discussion of how such mirrors work until Sec. 10.8.2). In Sec. 10.4, we also see how phase conjugation can be used to counteract the distortion of images and signals by media through which they travel.

In Sec. 10.5, we introduce nonlinear media and formulate Maxwell's equations for waves propagating through such media. As an example, we briefly discuss electro-optic effects, where a slowly changing electric field modulates the optical properties of a nonlinear crystal, thereby modulating light waves that propagate through it. Then in Sec. 10.6, we develop a detailed description of how such a nonlinear crystal couples two optical waves to produce a new, third wave—so-called "three-wave mixing." Three-wave mixing has many important applications in modern technology. In Sec. 10.7, we describe and analyze several: frequency doubling (e.g., in a green laser pointer), optical parametric amplification of signals, and driving light into a squeezed state (e.g., the squeezed vacuum of quantum electrodynamics).

In an isotropic medium, three-wave mixing is suppressed, but a new, fourth wave can be produced by three incoming waves. In Sec. 10.8, we describe and analyze this four-wave mixing and how it is used in phase-conjugate mirrors and produces problems in the optical fibers widely used to transmit internet, television, and telephone signals.

These topics just scratch the surface of the exciting field of nonlinear optics, but they will give the reader an overview and some major insights into this field, and into nonlinear phenomena in the physical world.

10.2 Lasers

10.2.1 Basic Principles of the Laser

In quantum mechanics one identifies three different types of interaction of light with material systems (atoms, molecules, atomic nuclei, electrons, etc.): (i) *spontaneous emission,* in which a material system in an excited state spontaneously drops into a state of lesser excitation and emits a photon in the process; (ii) *absorption,* in which an incoming photon is absorbed by a material system, exciting it; and (iii) *stimulated emission,* in which a material system, initially in some excited state, is "tickled" by passing photons, and this tickling stimulates it to emit a photon of the same sort (in the same state) as the photons that tickled it.

spontaneous emission, absorption, and stimulated emission

As peculiar as stimulated emission may seem at first sight, in fact it is easily understood and analyzed classically. It is nothing but "negative absorption": In classical physics, when a light beam with electric field $E = \Re[Ae^{i(kz-\omega t+\varphi)}]$ travels through an absorbing medium, its real amplitude A decays exponentially with the distance propagated, $A \propto e^{-\mu z/2}$ (corresponding to an energy-flux decay $F \propto e^{-\mu z}$), while its frequency ω, wave number k, and phase φ remain nearly constant. For normal materials, the absorption rate $\mu = F^{-1}dF/dz$ is positive, and the energy lost goes ultimately into heat. However, one can imagine a material with an internally stored energy that amplifies a passing light beam. Such a material would have a negative absorption rate, $\mu < 0$, and correspondingly, the amplitude of the passing light would grow with the distance traveled, $A \propto e^{+|\mu|z/2}$, while its frequency, wave number, and phase would remain nearly constant. Such materials do exist; they are called "active media," and their amplification of passing waves is stimulated emission.

active medium

This elementary, classical description of stimulated emission is equivalent to the quantum mechanical description in the domain where the stimulated emission is strong: the domain of large photon-occupation numbers $\eta \gg 1$ (which, as we learned in Sec. 3.2.5, is the domain of classical waves).

The classical description of stimulated emission takes for granted the existence of an active medium. To understand the nature of such a medium, we must turn to quantum mechanics.

As a first step toward such understanding, consider a beam of monochromatic light with frequency ω that impinges on a collection of molecules (or atoms or charged particles) that are all in the same quantum mechanical state $|1\rangle$. Suppose the molecules have a second state $|2\rangle$ with energy $E_2 = E_1 + \hbar\omega$. Then the light will resonantly excite the molecules from their initial state $|1\rangle$ to the higher state $|2\rangle$, and in the process photons will be absorbed (Fig. 10.1a). The strength of the interaction is proportional to the beam's energy flux F. Stated more precisely, the rate of absorption of photons

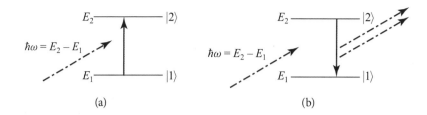

FIGURE 10.1 (a) *Photon absorption.* A photon with energy $\hbar\omega = E_2 - E_1$ excites a molecule from its ground state, with energy E_1 to an excited state with energy E_2 (as depicted by an energy-level diagram). (b) *Stimulated emission.* The molecule is initially in its excited state, and the incoming photon stimulates it to deexcite into its ground state, emitting a photon identical to the incoming one.

is proportional to the number flux of photons in the beam, $dn/dA\,dt = F/(\hbar\omega)$, and thence is proportional to F, in accord with the classical description of absorption.

Next suppose that when the light beam first arrives, the atoms are all in the higher state $|2\rangle$ rather than the lower state $|1\rangle$. There will still be a resonant interaction, but this time the interaction will deexcite the atoms, with an accompanying emission of photons (Fig. 10.1b). As in the absorption case, the strength of the interaction is proportional to the flux of the incoming beam (i.e., the rate of emission of new photons is proportional to the number flux of photons that the beam already has), and thence it is also proportional to the beam's energy flux F. A quantum mechanical analysis shows that the photons from this stimulated emission come out in the same quantum state as is occupied by the photons of the incoming beam (Bose-Einstein statistics: photons, being bosons, tend to congregate in the same state). Correspondingly, when viewed classically, the beam's flux will be amplified at a rate proportional to its initial flux, with no change of its frequency, wave number, or phase.

In Nature, molecules usually have their energy levels populated in accord with the laws of statistical (thermodynamic) equilibrium. Such thermalized populations, as we saw at the end of Sec. 4.4.1, entail a ratio $N_2/N_1 = \exp[-(E_2 - E_1)/(k_B T)] < 1$ for the number N_2 of molecules in state $|2\rangle$ to the number N_1 in state $|1\rangle$. Here T is the molecular temperature, and for simplicity it is assumed that the states are nondegenerate. Since there are more molecules in the lower state $|1\rangle$ than the higher one $|2\rangle$, an incoming light beam will experience more absorption than stimulated emission.

population inversion

By contrast, occasionally in Nature and often in the laboratory a collection of molecules develops a "population inversion" in which $N_2 > N_1$. The two states can then be thought of as having a negative temperature with respect to each other. Light propagating through population-inverted molecules will experience more stimulated emission than absorption (i.e., it will be amplified). The result is "light amplification by stimulated emission," or "laser" action.

laser

This basic principle underlying the laser has been known since the early years of quantum mechanics, but only in the 1950s did physicists succeed in designing, constructing, and operating real lasers. The first proposals for practical devices were

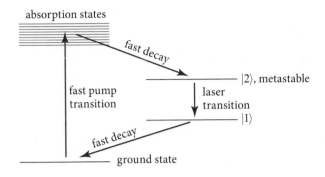

FIGURE 10.2 The mechanism for creating the population inversion that underlies laser action. The horizontal lines and band represent energy levels of a molecule, and the arrows represent transitions in which the molecules are excited by pumping or decay by emission of photons.

made, independently, in the United States by Weber (1953) and Gordon, Zeiger, and Townes (1954), and in the Soviet Union by Basov and Prokhorov (1954, 1955). The first successful construction and operation of a laser was by Gordon, Zeiger, and Townes (1954, 1955), and soon thereafter by Basov and Prokhorov—though these first lasers actually used radiation not at optical frequencies but rather at microwave frequencies and were based on a population inversion of ammonia molecules (see Feynman, Leighton, and Sands, 2013, Chap. 9), and thus were called *masers*. The first optical frequency laser, one based on a population inversion of chromium ions in a ruby crystal, was constructed and operated by Maiman (1960).

The key to laser action is the population inversion. Population inversions are incompatible with thermodynamic equilibrium; thus, to achieve them, one must manipulate the molecules in a nonequilibrium way. This is usually done by some concrete variant of the process shown in the energy-level diagram of Fig. 10.2. Some sort of pump mechanism (to be discussed in the next section) rapidly excites molecules from the ground state into some group of absorption states. The molecules then decay rapidly from the absorption states into the state $|2\rangle$, which is metastable (i.e., has a long lifetime against spontaneous decay), so the molecules linger there. The laser transition is from state $|2\rangle$ to state $|1\rangle$. Once a molecule has decayed into state $|1\rangle$, it quickly decays on down to the ground state and then may be quickly pumped back up into the absorption states. This is called "four-level pumping." It is instead called "three-level pumping" if state $|1\rangle$ is the ground state.

pump mechanism

If the pump acts suddenly and briefly, this process produces a temporary population inversion of states $|2\rangle$ and $|1\rangle$, with which an incoming, weak burst of "seed" light can interact to produce a burst of amplification. The result is a pulsed laser. If the pump acts continuously, the result may be a permanently maintained population inversion with which continuous seed light can interact to produce continuous-wave laser light.

pulsed lasers and continuous-wave lasers

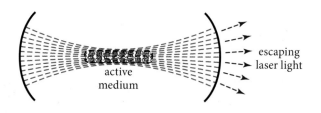

FIGURE 10.3 The use of a Fabry-Perot cavity to enhance the interaction of the light in a laser with its active medium.

As the laser beam travels through the active medium (the population-inverted molecules), its flux F builds up with distance z as $dF/dz = F/\ell_o$, so $F(z) = F_o e^{z/\ell_o}$. Here F_o is the initial flux, and $\ell_o \equiv 1/|\mu|$ (the e-folding length) depends on the strength of the population inversion and the strength of the coupling between the light and the active medium. Typically, ℓ_o is so long that strong lasing action cannot be achieved by a single pass through the active medium. In this case, the lasing action is enhanced by placing the active medium inside a Fabry-Perot cavity (Fig. 10.3 and Sec. 9.4.3). The length L of the cavity is adjusted to maximize the output power, which occurs when the lasing transition frequency $\omega = (E_2 - E_1)/\hbar$ is an eigenfrequency of the cavity. The lasing action then excites a standing wave mode of the cavity, from which the light leaks out through one or both cavity mirrors. If $\mathcal{F}$ is the cavity's finesse [approximately the average number of times a photon bounces back and forth inside the cavity before escaping through a mirror; cf. Eq. (9.39)], then the cavity increases the distance that typical photons travel through the active medium by a factor $\sim \mathcal{F}$, thereby increasing the energy flux of the light output by a factor $\sim e^{\mathcal{F} L/\ell_o}$.

Typically, many modes of the Fabry-Perot cavity are excited, so the laser's output is multimodal and contains a mixture of polarizations. When a single mode and polarization are desired, the polarization is made pure by oblique optical elements at the ends of the laser that transmit only the desired polarization, and all the modes except one are removed from the output light by a variety of techniques (e.g., filtering with a second Fabry-Perot cavity; Sec. 9.4.3).

coherent-state laser light

For an ideal laser (one, e.g., with a perfectly steady pump maintaining a perfectly steady population inversion that in turn maintains perfectly steady lasing), the light comes out in the most perfectly classical state that quantum mechanics allows. This state, called a *quantum mechanical coherent state*, has a perfectly sinusoidally oscillating electric field on which is superimposed the smallest amount of noise (the smallest wandering of phase and amplitude) allowed by quantum mechanics: the noise of quantum electrodynamical vacuum fluctuations. The value of the oscillations' well-defined phase is determined by the phase of the seed field from which the coherent state was built up by lasing. Real lasers have additional noise due to a variety of practical factors, but nevertheless, their outputs are usually highly coherent, with long coherence times.

10.2.2 Types of Lasers and Their Performances and Applications

Lasers can have continuous, nearly monochromatic output, or they can be pulsed. Their active media can be liquids, gases (ionized or neutral), or solids (semiconductors, glasses, or crystals; usually carefully doped with impurities). Lasers can be pumped by radiation (e.g., from a flash tube), by atomic collisions that drive the lasing atoms into their excited states, by nonequilibrium chemical reactions, or by electric fields associated with electric currents (e.g., in semiconductor diode lasers that can be powered by ordinary batteries and are easily modulated for optical communication).

laser pump mechanisms

Lasers can be made to pulse by turning the pump on and off, by mode-locked operation (Sec. 10.2.3), or by Q-switching (turning off the lasing, e.g., by inserting into the Fabry-Perot cavity an electro-optic material that absorbs light until the pump has produced a huge population inversion, and then suddenly applying an electric field to the absorber, which makes it transparent and restores the lasing).

pulsed lasers

Laser pulses can be as short as a few femtoseconds (thus enabling experimental investigations of fast chemical reactions) and they can carry as much as 20,000 J with duration of a few picoseconds and pulse power $\sim 10^{15}$ W (at the U.S. National Ignition Facility in Livermore, California, for controlled fusion).

The most powerful continuous laser in the United States is the Mid-Infrared Advanced Chemical Laser (MIRACL), developed by the Navy to shoot down missiles and satellites, with ~ 1 MW power in a 14×14 cm^2 beam lasting ~ 70 s. Continuous CO_2 lasers with powers ~ 3 kW are used industrially for cutting and welding metal.

continuous lasers

The beam from a Q-switched CO_2 laser with ~ 1 GW power can be concentrated into a region with transverse dimensions as small as one wavelength (~ 1 μm), yielding a local energy flux of 10^{21} W m^{-2}, an rms magnetic field strength of ~ 3 kT, an electric field of ~ 1 TV m^{-1}, and an electrical potential difference across a wavelength of ~ 1 MeV. It then should not be surprising that high-power lasers can create electron-positron-pair plasmas!

For many applications, large power is irrelevant or undesirable, but high frequency stability (a long coherence time) is often crucial. By locking the laser frequency to an optical frequency atomic transition (e.g., in the Al$^+$ atomic clock; see footnote in Sec. 6.6.1), one can achieve a frequency stability $\Delta f/f \sim 10^{-17}$, so $\Delta f \sim 3$ mHz, for hours or longer, corresponding to coherence times of ~ 100 s and coherence lengths of $\sim 3 \times 10^7$ km. By locking the frequency to a mode of a physically highly stable Fabry-Perot cavity (e.g., PDH locking; Sec. 9.4.3), stabilities have been achieved as high as $\Delta f/f \sim 10^{-16}$ for times of ~ 1 hr in a physically solid cavity (the superconducting cavity stabilized oscillator), and $\Delta f/f \sim 10^{-22}$ for a few ms in LIGO's 4-km-long cavity with freely hanging mirrors and sophisticated seismic isolation (Sec. 9.5).

locking a laser's frequency

When first invented, lasers were called "a solution looking for a problem." Now they permeate everyday life and high technology. Examples are supermarket bar-code readers, laser pointers, DVD players, eye surgery, laser printers, laser cutting and welding, laser gyroscopes (which are standard on commercial aircraft), laser-based surveying, Raman spectroscopy, laser fusion, optical communication, optically based computers, holography, maser amplifiers, and atomic clocks.

10.2.3 Ti:Sapphire Mode-Locked Laser

As a concrete example of a modern, specialized laser, consider the titanium-sapphire (Ti:sapphire) mode-locked laser (Cundiff, 2002) that is used to generate the optical frequency comb described in Sec. 9.4.3. Recall that this laser's light must be concentrated in a very short pulse that travels unchanged back and forth between its Fabry-Perot mirrors. The pulse is made from phase-locked (Gaussian) modes of the optical cavity that extend over a huge frequency band, $\Delta\omega \sim \omega$. Among other things, this mode-locked laser illustrates the use of an optical nonlinearity called the "Kerr effect," whose underlying physics we describe later in this chapter (Sec. 10.8.3).

As we discussed in Sec. 9.4.3, this mode-locked laser must (i) more strongly amplify modes with high energy flux than with low (this pushes the light into the short, high-flux pulse), and (ii) its (Gaussian) modes must have a group velocity V_g that is independent of frequency over the frequency band $\Delta\omega \sim \omega$ (this enables the pulse to stay short rather than disperse).

Figure 10.4 illustrates the Ti:sapphire laser that achieves this. The active medium is a sapphire crystal doped with titanium ions. This medium exhibits the optical Kerr effect, which means that its index of refraction n is a sum of two terms, one independent of the light's energy flux; the other proportional to the flux [Eq. (10.69)]. The flux-dependent term slows the light's speed near the beam's center and thereby focuses the beam, making its cross section smaller. A circular aperture attenuates large light beams but not small. As a result, the lasing is stronger the smaller the beam (which means the higher its flux). This drives the lasing light into the desired short, high-flux pulse.

The Ti:sapphire crystal has a group velocity that increases with frequency. The two prisms and mirror (Fig. 10.4) compensate this. The first prism bends low-frequency light more than high, so the high-frequency light traverses more glass in the second prism and is slowed. By adjusting the mirror tilt, one adjusts the amount of slowing to keep the round-trip-averaged phase velocity the same at high frequencies as at low.

mode-locked laser produces frequency comb

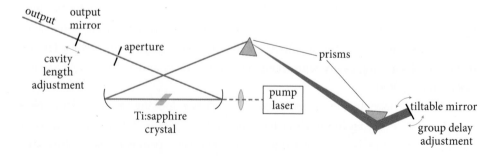

FIGURE 10.4 The Ti:sapphire mode-locked laser. Adapted from Cundiff (2002).

The laser then generates multimode, high-intensity, pulsed light (with pulse durations that can be as short as attoseconds, when augmented by a technique called chirped pulse amplification), resulting in the optical frequency comb of Sec. 9.4.3.

10.2.4 Free Electron Laser

The free electron laser is quite different from the lasers we have been discussing. A collimated beam of essentially monoenergetic electrons (the population inversion) passes through an undulator comprising an alternating transverse magnetic field, which causes the electrons to radiate X-rays along the direction of the beam. From a classical perspective, this X-ray-frequency electromagnetic wave induces correlated electron motions, which emit coherently, enhancing the X-ray wave. In other words, the undulating electron beam lases.

As an example, the LINAC Coherent Light Source (LCLS) at SLAC National Accelerator Laboratory can produce femtosecond pulses of keV X-rays with pulse frequency 120 Hz and peak intensity $\sim 10^9$ times higher than that achievable from a conventional synchrotron. An upgrade, LCLS-II, is planned to start operating in 2020 with a pulse frequency ~ 100 kHz and an increase in average brightness of $\sim 10^3$.

EXERCISES

Exercise 10.1 *Challenge: Nuclear Powered X-Ray Laser*
A device much ballyhooed in the United States during the presidency of Ronald Reagan, but thankfully never fully deployed, was a futuristic, superpowerful X-ray laser pumped by a nuclear explosion. As part of Reagan's Strategic Defense Initiative ("Star Wars"), this laser was supposed to shoot down Soviet missiles.

How would you design a nuclear powered X-ray laser? The energy for the pump comes from a nuclear explosion that you set off in space above Earth. You want to use that energy to create a population inversion in an active medium that will lase at X-ray wavelengths, and you want to focus the resulting X-ray beam onto an intercontinental ballistic missile that is rising out of Earth's atmosphere. What would you use for the active medium? How would you guarantee that a population inversion is created in the active medium? How would you focus the resulting X-ray beam? (Note: This is a highly nontrivial exercise, intended more as a stimulus for thought than as a test of one's understanding of things taught in this book.)

10.3 Holography

Holography is an old[1] and well-explored example of nonlinear optics—an example in which the nonlinear interaction of light with itself is produced not in real time, but rather by means of a recording followed by a later readout (e.g., Cathey, 1974; Iizuka, 1987).

---

1. Holography was invented by Dennis Gabor (1948) for use in electron microscopy. The first practical optical holograms were made in 1962, soon after lasers became available.

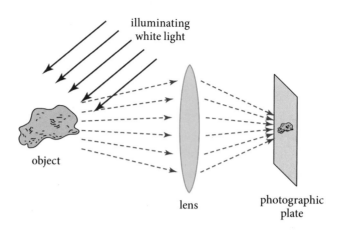

FIGURE 10.5 Ordinary photography.

By contrast with ordinary photography (Fig. 10.5), which produces a colored, 2-dimensional image of 3-dimensional objects, holography (Figs. 10.6 and 10.8) normally produces a monochromatic 3-dimensional image of 3- dimensional objects. Roughly speaking, the two processes contain the same amount of information, two items at each location in the image. For ordinary photography, they are the energy flux and color; for holography, the energy flux and phase of monochromatic light.

It is the phase, lost from an ordinary photograph but preserved in holography, that carries the information about the third dimension. Our brain deduces the distance to a point on an object from the difference in the directions of propagation of the point's light as it arrives at our two eyes. Those propagation directions are encoded in the light as variations of the phase with transverse location [see, e.g., the point-spread function for a thin lens, Eq. (8.29)].

ordinary photography

In an ordinary photograph (Fig. 10.5), white light scatters off an object, with different colors scattering at different strengths. The resulting colored light is focused through a lens to form a colored image on a photographic plate or a CCD. The plate or CCD records the color and energy flux at each point or pixel in the focal plane, thereby producing the ordinary photograph.

hologram

In holography, one records a hologram with flux and phase information (see Fig. 10.6), and one then uses the hologram to reconstruct the 3-dimensional, monochromatic, holographic image (see Fig. 10.8).

10.3.1

10.3.1 Recording a Hologram

Consider, first, the recording of the hologram. Monochromatic, linearly polarized plane-wave light with electric field

$$E = \Re[\psi(x, y, z)e^{-i\omega t}], \tag{10.1}$$

angular frequency ω, and wave number $k = \omega/c$ illuminates the object and also a mirror, as shown in Fig. 10.6. The light must be spatially coherent over the entire

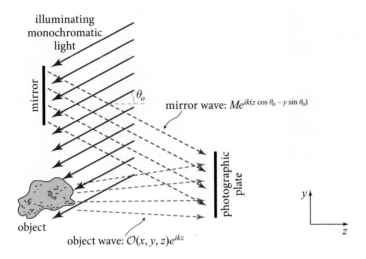

illuminating monochromatic light

mirror

θ_o

mirror wave: $Me^{ik(z\cos\theta_o - y\sin\theta_o)}$

photographic plate

object

object wave: $\mathcal{O}(x, y, z)e^{ikz}$

y

z

FIGURE 10.6 Recording a hologram.

region of mirror plus object. The propagation vector **k** of the illuminating light lies in the y-z plane at some angle θ_o to the z-axis, and the mirror lies in the x-y plane. The mirror reflects the illuminating light, producing a so-called "reference beam," which we call the *mirror wave:*

reference beam or mirror wave

$$\psi_{\text{mirror}} = Me^{ik(z\cos\theta_o - y\sin\theta_o)}, \tag{10.2}$$

where M is a real constant. The object (shown in Fig. 10.6) scatters the illuminating light, producing a wave that propagates in the z direction toward the recording medium (a photographic plate, for concreteness). We call this the *object wave* and denote it

object wave

$$\psi_{\text{object}} = \mathcal{O}(x, y, z)e^{ikz}. \tag{10.3}$$

It is the slowly varying complex amplitude $\mathcal{O}(x, y, z)$ of this object wave that carries the 3-dimensional, but monochromatic, information about the object's appearance. Thus it is this $\mathcal{O}(x, y, z)$ that will be reconstructed in the second step of holography.

In the first step (Fig. 10.6), the object wave propagates along the z direction to the photographic plate (or, more commonly today, a photoresist) at $z = 0$, where it interferes with the mirror wave to produce the transverse pattern of energy flux

$$F(x, y) \propto |\mathcal{O} + Me^{-iky\sin\theta_o}|^2$$
$$= M^2 + |\mathcal{O}(x, y, z = 0)|^2 + \mathcal{O}(x, y, z = 0)Me^{iky\sin\theta_o}$$
$$+ \mathcal{O}^*(x, y, z = 0)Me^{-iky\sin\theta_o}. \tag{10.4}$$

(Here and throughout this chapter, * denotes complex conjugation.) The plate is blackened at each point in proportion to this flux. The plate is then developed, and a positive or negative print (it doesn't matter which because of Babinet's principle,

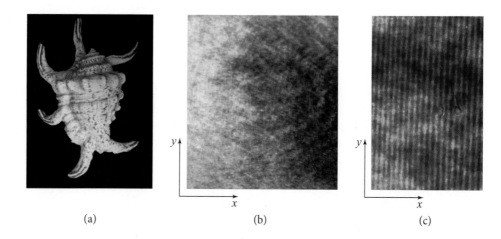

FIGURE 10.7 (a) Ordinary photograph of an object. (b) Hologram of the same object. (c) Magnification of the hologram. Photos courtesy Jason Sapan, Holographic Studios, New York City.

Sec. 8.3.3) is made on a transparent sheet of plastic or glass. This print, the *hologram*, has a transmissivity as a function of x and y that is proportional to the flux distribution (10.4):

hologram's transmissivity

$$\mathfrak{t}(x, y) \propto M^2 + |\mathcal{O}(x, y, z = 0)|^2 + \mathcal{O}(x, y, z = 0)Me^{iky\sin\theta_o}$$
$$+ \mathcal{O}^*(x, y, z = 0)Me^{-iky\sin\theta_o}. \tag{10.5}$$

hologram's nonlinearity

In this transmissivity we meet our first example of nonlinearity: $\mathfrak{t}(x, y)$ contains a nonlinear superposition of the mirror wave and the object wave. Stated more precisely, the superposition is not a linear sum of wave fields, but instead is a sum of products of one wave field with the complex conjugate of another wave field. A further nonlinearity will arise in the reconstruction of the holographic image [Eq. (10.7)].

Figure 10.7 shows an example. Figure 10.7a is an ordinary photograph of an object, Fig. 10.7b is a hologram of the same object, and Fig. 10.7c is a blow-up of a portion of that hologram. The object is not at all recognizable in the hologram, because the object wave $\mathcal{O}$ was not focused to form an image at the plane of the photographic plate. Rather, light from each region of the object was scattered to and recorded by all regions of the photographic plate. Nevertheless, the plate contains the full details of

hologram's information encoding

the scattered light $\mathcal{O}(x, y, z = 0)$, including its phase. That information is recorded in the piece $M(\mathcal{O}e^{iky\sin\theta_o} + \mathcal{O}^*e^{-iky\sin\theta_o}) = 2M\,\Re(\mathcal{O}e^{iky\sin\theta_o})$ of the hologram's transmissivity. This piece oscillates sinusoidally in the y direction with wavelength $2\pi/(k\sin\theta_o)$; and the amplitude and phase of its oscillations are modulated by the object wave $\mathcal{O}(x, y, z = 0)$. Those modulated oscillations show up clearly when one magnifies the hologram (Fig. 10.7c); they turn the hologram into a sort of diffraction grating, with the object wave $\mathcal{O}(x, y, z = 0)$ encoded as variations of the darkness and spacings of the grating lines.

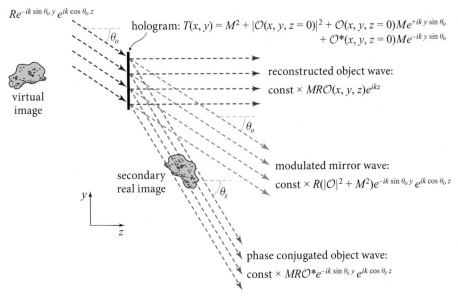

reference wave:

$Re^{-ik\sin\theta_o\,y}\,e^{ik\cos\theta_o\,z}$

hologram: $T(x, y) = M^2 + |\mathcal{O}(x, y, z = 0)|^2 + \mathcal{O}(x, y, z = 0)Me^{+ik\,y\sin\theta_o}$
$+ \mathcal{O}^*(x, y, z = 0)Me^{-ik\,y\sin\theta_o}$

reconstructed object wave:

const $\times MR\mathcal{O}(x, y, z)e^{ikz}$

virtual image

modulated mirror wave:

const $\times R(|\mathcal{O}|^2 + M^2)e^{-ik\sin\theta_o\,y}\,e^{ik\cos\theta_o\,z}$

secondary real image

phase conjugated object wave:

const $\times MR\mathcal{O}^*e^{-ik\sin\theta_s\,y}\,e^{ik\cos\theta_s\,z}$

FIGURE 10.8 Reconstructing the holographic image from the hologram. Note that $\sin\theta_s = 2\sin\theta_o$.

What about the other pieces of the transmissivity (10.5), which superpose linearly on the diffraction grating? One piece, $\mathfrak{t} \propto M^2$, is spatially uniform and thus has no effect except to make the lightest parts of the hologram slightly gray rather than leaving it absolutely transparent (since this hologram is a negative rather than a positive). The other piece, $\mathfrak{t} \propto |\mathcal{O}|^2$, is the flux of the object's unfocused, scattered light. It produces a graying and whitening of the hologram (Fig. 10.7b) that varies on lengthscales long compared to the grating's wavelength $2\pi/(k\sin\theta_o)$ and thus blots out the diffraction grating a bit here and there, but it does not change the amplitude or phase of the grating's modulation.

10.3.2 Reconstructing the 3-Dimensional Image from a Hologram

10.3.2

To reconstruct the object's 3-dimensional wave, $\mathcal{O}(x, y, z)e^{ikz}$, one sends through the hologram monochromatic, plane-wave light identical to the mirror light used in making the hologram (Fig. 10.8). If, for pedagogical simplicity, we place the hologram at the same location $z = 0$ as was previously occupied by the photographic plate, then the incoming light has the same form (10.2) as the original mirror wave, but with an amplitude that we denote as R, corresponding to the phrase *reference beam* that is used to describe this incoming light:

reference beam for reconstructing 3-dimensional image

$$\psi_{\text{reference}} = Re^{ik(z\cos\theta_o - y\sin\theta_o)}. \qquad (10.6)$$

In passing through the hologram at $z = 0$, this reference beam is partially absorbed and partially transmitted. The result, immediately on exiting from the hologram, is a reconstructed light-wave field whose value and normal derivative are given by [cf. Eq. (10.5)]

reconstructed wave

$$\psi_{\text{reconstructed}}\Big|_{z=0} \equiv \mathcal{R}(x, y, z=0) = \mathfrak{t}(x, y)Re^{-iky\sin\theta_o}$$

$$= \left[M^2 + |\mathcal{O}(x, y, z=0)|^2\right]Re^{-iky\sin\theta_o}$$

$$+ MR\mathcal{O}(x, y, z=0)$$

$$+ MR\mathcal{O}^*(x, y, z=0)e^{-i2ky\sin\theta_o};$$

$$\psi_{\text{reconstructed},z}\Big|_{z=0} \equiv \mathcal{Z}(x, y, z=0) = ik\cos\theta_o\mathcal{R}(x, y, z=0). \tag{10.7}$$

This field and normal derivative act as initial data for the subsequent evolution of the reconstructed wave. Note that the field and derivative, and thus also the reconstructed wave, are triply nonlinear: each term in Eq. (10.7) is a product of (i) the original mirror wave M used to construct the hologram or the original object wave $\mathcal{O}$, (ii) $\mathcal{O}^*$ or $M^* = M$, and (iii) the reference wave R that is being used in the holographic reconstruction.

The evolution of the reconstructed wave beyond the hologram (at $z > 0$) can be computed by combining the initial data [Eq. (10.7)] for $\psi_{\text{reconstructed}}$ and $\psi_{\text{reconstructed},z}$ at $z = 0$ with the Helmholtz-Kirchhoff formula (8.4); see Exs. 10.2 and 10.6. From the four terms in Eq. (10.7) [which arise from the four terms in the hologram's transmissivity $\mathfrak{t}(x, y)$, Eq. (10.5)], the reconstruction produces four wave fields; see Fig. 10.8. The direction of propagation of each of these waves can easily be inferred from the vertical spacing of its phase fronts along the outgoing face of the hologram, or equivalently, from the relation $\partial\psi_{\text{reconstructed}}/\partial y = ik_y\psi_{\text{reconstructed}} = -ik\sin\theta\psi$, where θ is the angle of propagation relative to the horizontal z direction. Since, immediately in front of the hologram, $\psi_{\text{reconstructed}} = \mathcal{R}$, the propagation angle is

$$\sin\theta = \frac{\partial\mathcal{R}/\partial y}{-ik\mathcal{R}}. \tag{10.8}$$

Comparing with Eqs. (10.5) and (10.7), we see that the first two, slowly spatially varying, terms in the transmissivity, $\mathfrak{t} \propto M^2$ and $T \propto |\mathcal{O}|^2$, both produce waves that propagate in the same direction as the reference wave, $\theta = \theta_o$. This combined wave has an uninteresting, smoothly and slowly varying energy-flux pattern.

The two diffraction-grating terms in the hologram's transmissivity produce two interesting waves. One, arising from $\mathfrak{t} \propto \mathcal{O}(x, y, z=0)Me^{iky\sin\theta_o}$ [and produced by the $MR\mathcal{O}$ term of the initial conditions (10.7)], is *precisely the same object wave* $\psi_{\text{object}} = \mathcal{O}(x, y, z)e^{ikz}$ (aside from overall amplitude) *as one would have seen while making the hologram if one had replaced the photographic plate by a window and looked through it*. This object wave, carrying [encoded in $\mathcal{O}(x, y, z)$] the holographic image with full 3-dimensionality, propagates in the z direction, $\theta = 0$.

object wave and its 3-dimensional image

The transmissivity's second diffraction-grating term,

secondary (phase conjugate) wave

$$\mathfrak{t} \propto \mathcal{O}^*(x, y, z = 0) M e^{-iky \sin \theta_o},$$

acting via the $M R \mathcal{O}^*$ term of the initial conditions (10.7), gives rise to a secondary wave, which [according to Eq. (10.8)] propagates at an angle θ_s to the z-axis, where

$$\sin \theta_s = 2 \sin \theta_o. \tag{10.9}$$

(If $\theta_o > 30°$, then $2 \sin \theta_o > 1$, which means θ_s cannot be a real angle, and there will be no secondary wave.) *This secondary wave, if it exists, carries an image that is encoded in the complex conjugate $\mathcal{O}^*(x, y, z = 0)$ of the transverse (i.e., x, y) part of the original object wave.* Since complex conjugation of an oscillatory wave just reverses the sign of the wave's phase, this wave in some sense is a "phase conjugate" of the original object wave.

When one recalls that the electric and magnetic fields that make up an electromagnetic wave are actually real rather than complex, and that we are using complex wave fields to describe electromagnetic waves only for mathematical convenience, one then realizes that this phase conjugation of the object wave is actually a highly nonlinear process. There is no way, by linear manipulations of the real electric and magnetic fields, to produce the phase-conjugated wave from the original object wave.

In Sec. 10.4 we develop in detail the theory of phase-conjugated waves, and in Ex. 10.6, we relate our holographically constructed secondary wave to that theory. As we shall see, our secondary wave is not quite the same as the phase-conjugated object wave, but it is the same aside from some distortion along the y direction and a change in propagation direction. More specifically, *if one looks into the object wave with one's eyes (i.e., if one focuses it onto one's retinas), one sees the original object in all its 3-dimensional glory, though single colored, sitting behind the hologram at the object's original position* (shown in Fig. 10.8). Because the image one sees is behind the hologram, it is called a *virtual image. If, instead, one looks into the secondary wave with one's eyes, one sees the original 3-dimensional object, sitting in front of the hologram but turned inside out and distorted* (also shown in the figure). For example, if the object is a human face, the secondary image looks like the interior of a mask made from that human face, with distortion along the y direction. Because this secondary image appears to be in front of the hologram, it is called a *real image*—even though one can pass one's hands through it and feel nothing but thin air.

object wave's image is virtual

secondary wave's image is real

10.3.3 Other Types of Holography; Applications

10.3.3

There are many variants on the basic holographic technique depicted in Figs. 10.6–10.8. These include the following.

PHASE HOLOGRAPHY

Instead of darkening the high-flux regions of the hologram as in photography, one produces a phase-shifting screen, whose phase shift (due to thickening of the hologram's material) is proportional to the incoming flux. Such a phase hologram transmits more of the reference-wave light than a standard, darkened hologram, thus making a brighter reconstructed image.

VOLUME HOLOGRAPHY

The hologram is a number of wavelengths deep rather than being just 2-dimensional. For example, it could be made from a thick photographic emulsion, in which the absorption length for light is longer than the thickness. Such a hologram has a 3-dimensional grating structure (grating "surfaces" rather than grating "lines"), with two consequences. When one reconstructs the holographic image from it in the manner of Fig. 10.8, (i) the third dimension of the grating suppresses the secondary wave while enhancing the (desired) object wave, so more power goes into it; and (ii) the reference wave's incoming angle θ_o must be controlled much more precisely, as modest errors suppress the reconstructed object wave. This second consequence enables one to record multiple images in a volume hologram, each using its own angle θ_o for the illuminating light and reference wave.

3-dimensional gratings

REFLECTION HOLOGRAPHY

One reads out the hologram by reflecting light off of it rather than transmitting light through it, and the hologram's diffraction grating produces a 3-dimensional holographic image by the same process as in transmission; see Ex. 10.3.

WHITE-LIGHT HOLOGRAPHY

The hologram is recorded with monochromatic light as usual, but it is optimized for reading out with white light. Even for the simple 2-dimensional hologram of Fig. 10.8, if one sends in white light at the angle θ_o, one will get a 3-dimensional object wave: the hologram's grating will diffract various wavelengths in various directions. In the direction of the original object wave (the horizontal direction in Fig. 10.8), one gets a 3-dimensional reconstructed image of the same color as was used when constructing the hologram. When one moves away from that direction (vertically in Fig. 10.8), one sees the color of the 3-dimensional image continuously change (Ex. 10.2b). White-light reflection holograms are used on credit cards and money as impediments to counterfeiting; they have even been used on postage stamps.

colored holograms

COMPUTATIONAL HOLOGRAMS

Just as one can draw 2-dimensional pictures numerically, pixel-by-pixel, so one can also create and modify holograms numerically, then read them out optically.

FULL-COLOR HOLOGRAPHY

A full-color holographic image of an object can be constructed by superposing three monochromatic holographic images with the three primary colors—red, green, and

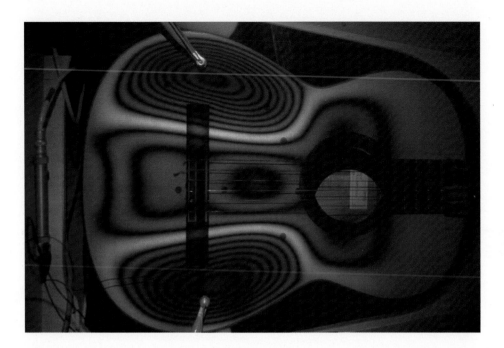

FIGURE 10.9 In-and-out vibrations of a guitar body visualized via holographic interferometry with green light. The dark and bright curves are a contour map, in units of the green light's wavelength, of the amplitude of vibration. Image courtesy Bernard Richardson, Cardiff University.

blue. One way to achieve this is to construct a single volume hologram using illuminating light from red, green, and blue laser beams, each arriving from a different 2-dimensional direction θ_o. Each beam produces a diffraction grating in the hologram with a different orientation and with spatial wave number corresponding to the beam's color. The 3-dimensional image can then be reconstructed using three white-light reference waves, one from each of the original three directions θ_o. The hologram will pick out of each beam the appropriate primary color, and produce the desired three overlapping images, which the eye will interpret as having approximately the true colors of the original object.

combining red, green, and blue holograms

HOLOGRAPHIC INTERFEROMETRY

One can observe changes in the shape of a surface at about the micrometer level by constructing two holograms, one of the original surface and the other of the changed surface, and then interfering the reconstructed light from the two holograms. Among other things, this holographic interferometry is used to observe small strains and vibrations of solid bodies—for example, sonic vibrations of a guitar in Fig. 10.9.

HOLOGRAPHIC LENSES

Instead of designing a hologram to reconstruct a 3-dimensional image, one can design it to manipulate light beams in most any way one wishes. As a simple example (Ex. 10.4c) of such a holographic lens, one can construct a holographic lens that splits one beam into two and focuses each of the two beams on a different spot. Holographic lenses are widely used in everyday technology, for example, to read bar codes in supermarket checkouts and to read the information off CDs, DVDs, and BDs (Ex. 10.5).

FUTURE APPLICATIONS

Major applications of holography that are under development include (i) dynamically changing volume holograms for 3-dimensional movies (which, of course, will require no eye glasses), and (ii) volume holograms for storage of large amounts of data—up to terabytes cm^{-3}.

EXERCISES

Exercise 10.2 *Derivation and Problem: Holographically Reconstructed Wave*

(a) Use the Helmholtz-Kirchhoff integral (8.4) or (8.6) to compute all four pieces of the holographically reconstructed wave field. Show that the piece generated by

$$\mathfrak{t} \propto \mathcal{O}(x, y, z = 0) M e^{iky \sin \theta_o}$$

is the same (aside from overall amplitude) as the field $\psi_{object} = \mathcal{O}(x, y, z)e^{-i\omega t}$ that would have resulted if when making the hologram (Fig. 10.6), the mirror wave had been absent and the photographic plate replaced by a window. Show that the other pieces have the forms and propagation directions indicated heuristically in Fig. 10.8. We shall examine the secondary wave, generated by $\mathfrak{t} \propto M\mathcal{O}^* e^{-iky \sin \theta_o}$, in Ex. 10.6.

(b) Suppose that plane-parallel white light is used in the holographic reconstruction of Fig. 10.8. Derive an expression for the direction in which one sees the object's 3-dimensional image have a given color (or equivalently, wave number). Assume that the original hologram was made with green light and $\theta_o = 45°$. What are the angles at which one sees the image as violet, green, yellow, and red?

Exercise 10.3 *Problem: Recording a Reflection Hologram*

How would you record a hologram if you want to read it out via reflection? Draw diagrams illustrating this, similar to Figs. 10.6 and 10.8. [Hint: The mirror wave and object wave can impinge on the photographic plate from either side; it's your choice.]

Exercise 10.4 *Example: Holographic Lens to Split and Focus a Light Beam*

A holographic lens, like any other hologram, can be described by its transmissivity $\mathfrak{t}(x, y)$.

(a) What $\mathfrak{t}(x, y)$ will take a reference wave, impinging from the θ_o direction (as in Fig. 10.8) and produce from it a primary object wave that converges on the spot

$(x, y, z) = (0, 0, d)$? [Hint: Consider, at the hologram's plane, a superposition of the incoming mirror wave and the point-spread function (8.28), which represents a beam that diverges from a point source. Then phase conjugate the point-spread function, so it converges to a point instead of diverging.]

(b) Draw a contour plot of the transmissivity $t(x, y)$ of the lens in part (a). Notice the resemblance to the Fresnel zone plate of Sec. 8.4.4. Explain the connection of the two, paying attention to how the holographic lens changes when one alters the chosen angle θ_o of the reference wave.

(c) What $t(x, y)$ will take a reference wave, impinging from the θ_o direction, and produce from it a primary wave that splits in two, with equal light powers converging on the spots $(x, y, z) = (-a, 0, d)$ and $(x, y, z) = (+a, 0, d)$?

Exercise 10.5   **Problem: CDs, DVDs, and BDs*
Information on CDs, DVDs, and BDs (compact, digital video, and blu-ray disks) is recorded and read out using holographic lenses, but it is not stored holographically. Rather, it is stored in a linear binary code consisting of pits and no-pits (for 0 and 1) along a narrow spiraling track. In each successive generation of storage device, the laser light has been pushed to a shorter wavelength ($\lambda = 780$ nm for CDs, 650 nm for DVDs, and 405 nm for BDs), and in each generation, the efficiency of the information storage has been improved. In CDs, the information is stored in a single holographic layer on the surface of the disk; in DVDs and BDs, it is usually stored in a single layer but can also be stored in as many as four layers, one above the other, though with a modest price in access time.

(a) Explain why one can expect to record in a disk's recording layer, at the very most, (close to) four bits of information per square wavelength of the recording light.

(b) The actual storage capacities are up to 900 MB for CDs, 4.7 GB for DVDs, and 25 GB for BDs. How efficient are each of these technologies relative to the maximum given in part (a)?

(c) Estimate the number of volumes of the *Encyclopedia Britannica* that can be recorded on a CD, on a DVD, and on a BD.

10.4 Phase-Conjugate Optics

Nonlinear optical techniques make it possible to phase conjugate an optical wave in real time, by contrast with holography, where the phase conjugation requires recording a hologram and then reconstructing the wave later. In this section, we explore the properties of phase-conjugated waves of any sort (light, sound, plasma waves, etc.), and in the next section, we discuss technology by which real-time phase conjugation is achieved for light.

The basic ideas and foundations for phase conjugation of waves were laid in Moscow by Boris Yakovlevich Zel'dovich[2] and his colleagues (1972) and at Caltech by Amnon Yariv (1978).

Phase conjugation is the process of taking a monochromatic wave

$$\Psi_O = \Re[\psi(x, y, z)e^{-i\omega t}] = \frac{1}{2}(\psi e^{-i\omega t} + \psi^* e^{+i\omega t}), \tag{10.10a}$$

and from it constructing the wave

$$\Psi_{PC} = \Re[\psi^*(x, y, z)e^{-i\omega t}] = \frac{1}{2}(\psi^* e^{-i\omega t} + \psi e^{+i\omega t}). \tag{10.10b}$$

Notice that the phase-conjugated wave Ψ_{PC} is obtainable from the original wave Ψ_O by time reversal, $t \to -t$. This has a number of important consequences. One is that Ψ_{PC} propagates in the opposite direction to Ψ_O. Others are explained most clearly with the help of a *phase-conjugating mirror*.

Consider a wave Ψ_O with spatial modulation (i.e., a wave that carries a picture or a signal of some sort). Let the wave propagate in the z direction (rightward in Fig. 10.10), so that

$$\psi = \mathcal{A}(x, y, z)e^{i(kz-\omega t)}, \tag{10.11}$$

where $\mathcal{A} = A e^{i\varphi}$ is a complex amplitude whose modulus A and phase φ change slowly in x, y, and z (slowly compared to the wave's wavelength $\lambda = 2\pi/k$). Suppose that this wave propagates through a time-independent medium with slowly varying physical properties [e.g., a dielectric medium with slowly varying index of refraction $\mathfrak{n}(x, y, z)$]. These slow variations will distort the wave's complex amplitude as it propagates. The wave equation for the real, classical field $\Psi = \Re[\psi e^{-i\omega t}]$ will have the form $\mathcal{L}\Psi - \partial^2\Psi/\partial t^2 = 0$, where $\mathcal{L}$ is a real spatial differential operator that depends on the medium's slowly varying physical properties. This wave equation implies that the complex field ψ satisfies

$$\mathcal{L}\psi + \omega^2\psi = 0 \tag{10.12}$$

[which is the Helmholtz equation (8.1b) if $\mathcal{L}$ is the vacuum wave operator]. Equation (10.12) is the evolution equation for the wave's complex amplitude.

Let the distorted, rightward propagating wave Ψ_O reflect off a mirror located at $z = 0$. If the mirror is a phase-conjugating one, then very near it (at z near zero) the reflected wave will have the form

$$\Psi_{PC} = \Re[\mathcal{A}^*(x, y, z = 0)e^{i(-kz-\omega t)}], \tag{10.13}$$

---

2. Zel'dovich is the famous son of a famous Russian/Jewish physicist, Yakov Borisovich Zel'dovich, who with Andrei Dmitrievich Sakharov fathered the Soviet hydrogen bomb and then went on to become a dominant figure internationally in astrophysics and cosmology.

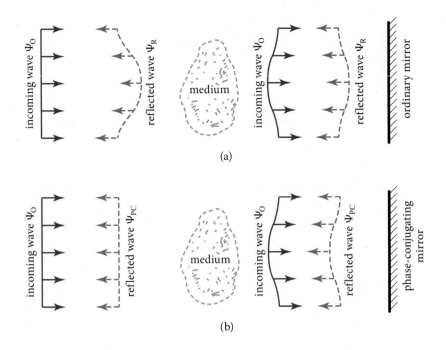

FIGURE 10.10 A rightward propagating wave and the reflected wave produced by (a) an ordinary mirror and (b) a phase-conjugating mirror. In both cases the waves propagate through a medium with spatially variable properties, which distorts their phase fronts. In case (a) the distortion is reinforced by the second passage through the variable medium; in case (b) the distortion is removed by the second passage.

while if it is an ordinary mirror, then the reflected wave will be

ordinary mirror and its reflected wave Ψ_R

$$\Psi_R = \Re[\pm \mathcal{A}(x, y, z = 0)e^{i(-kz-\omega t)}]. \tag{10.14}$$

(Here the sign depends on the physics of the wave. For example, if Ψ is the transverse electric field of an electromagnetic wave and the mirror is a perfect conductor, the sign will be minus to guarantee that the total electric field, original plus reflected, vanishes at the mirror's surface.)

These two waves, the phase-conjugated one Ψ_{PC} and the ordinary reflected one Ψ_R, have very different surfaces of constant phase (*phase fronts*). The phase of the incoming wave Ψ_O [Eqs. (10.10a) and (10.11)] as it nears the mirror ($z = 0$) is $\varphi + kz$, so (taking account of the fact that φ is slowly varying) the surfaces of constant phase are $z = -\varphi(x, y, z = 0)/k$. Similarly, the phase of the wave Ψ_R [Eq. (10.14)] reflected from the ordinary mirror is $\varphi - kz$, so its surfaces of constant phase near the mirror are $z = +\varphi(x, y, z = 0)/k$, which are reversed from those of the incoming wave, as shown in the upper right of Fig. 10.10. Finally, the phase of the wave Ψ_{PC} [Eq. (10.13)] reflected from the phase-conjugating mirror is $-\varphi - kz$, so its surfaces of constant phase near the mirror are $z = -\varphi(x, y, z = 0)/k$, which are the same as those of the

phase fronts of Ψ_{PC} and Ψ_R

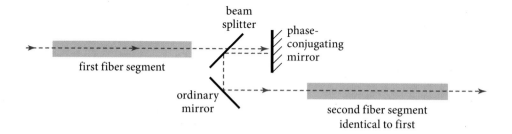

FIGURE 10.11 The use of a phase-conjugating mirror in an optical transmission line to prevent the fiber from distorting an optical image. The distortions put onto the image as it propagates through the first segment of fiber are removed during propagation through the second segment.

incoming wave (lower right of Fig. 10.10), even though the two waves are propagating in opposite directions.

The phase fronts of the original incoming wave and the phase-conjugated wave are the same not only near the phase-conjugating mirror; they are the same everywhere. More specifically, as the phase-conjugated wave Ψ_{PC} propagates away from the mirror [near which it is described by Eq. (10.13)], the propagation equation (10.12) forces it to evolve in such a way as to remain always the phase conjugate of the incoming wave:

$$\Psi_{PC} = \Re[\mathcal{A}^*(x, y, z)e^{-ikz}e^{-i\omega t}]. \tag{10.15}$$

This should be obvious: because the differential operator $\mathcal{L}$ in the propagation equation (10.12) for $\psi(x, y, z) = \mathcal{A}e^{ikz}$ is real, $\psi^*(x, y, z) = \mathcal{A}^*e^{-ikz}$ will satisfy this propagation equation when $\psi(x, y, z)$ does.

distortion removal in Ψ_{PC}

That the reflected wave Ψ_{PC} always remains the phase conjugate of the incoming wave Ψ_O means that *the distortions put onto the incoming wave, as it propagates rightward through the inhomogeneous medium, get removed from the phase-conjugated wave as it propagates back leftward* (Fig. 10.10).

This removal of distortions has a number of important applications. One is for image transmission in optical fibers. Normally when an optical fiber is used to transmit an optical image, the transverse spatial variations $n(x, y)$ of the fiber's index of refraction (which are required to hold the light in the fiber; Ex. 7.8) distort the image somewhat. The distortions can be eliminated by using a sequence of identical segments of optical fibers separated by phase-conjugating mirrors (Fig. 10.11). A few other applications include (i) real-time holography; (ii) removal of phase distortions in Fabry-Perot cavities by making one of the mirrors a phase-conjugating one, with a resulting improvement in the shape of the beam that emerges from the cavity; (iii) devices that can memorize an optical image and compare it to other images; (iv) the production of squeezed light (Ex. 10.16); and (v) improved focusing of laser light for laser fusion.

As we shall see in the next section, phase-conjugating mirrors rely crucially on the sinusoidal time evolution of the wave field; they integrate that sinusoidal evolution coherently over some timescale $\hat{\tau}$ (typically microseconds to nanoseconds) to produce the phase-conjugated wave. Correspondingly, if an incoming wave varies on timescales τ long compared to this $\hat{\tau}$ (e.g., if it carries a temporal modulation with bandwidth $\Delta\omega \sim 1/\tau$ small compared to $1/\hat{\tau}$), then the wave's temporal modulations will *not* be time reversed by the phase-conjugating mirror. For example, if the wave impinging on a phase-conjugating mirror has a frequency that is ω_a initially, and then gradually, over a time τ, increases to $\omega_b = \omega_a + 2\pi/\tau$, then the phase-conjugated wave will not emerge from the mirror with frequency ω_b first and ω_a later. Rather, it will emerge with ω_a first and ω_b later (same order as for the original wave). When the incoming wave's temporal variations are fast compared to the mirror's integration time, $\tau \ll \hat{\tau}$, the mirror encounters a variety of frequencies during its integration time and ceases to function properly. Thus, *even though phase conjugation is equivalent to time reversal in a formal sense, a phase-conjugating mirror cannot time reverse a temporal signal. It only time reverses monochromatic waves (which might carry a spatial signal).*

phase conjugation and time reversal

Exercise 10.6 *Derivation and Example: Secondary Wave in Holography*
Consider the secondary wave generated by $t \propto M\mathcal{O}^* e^{-iky\sin\theta_o}$ in the holographic reconstruction process of Fig. 10.8, Eq. (10.7), and Ex. 10.2.

(a) Assume, for simplicity, that the mirror and reference waves propagate nearly perpendicular to the hologram, so $\theta_o \ll 90°$ and $\theta_s \simeq 2\theta_o \ll 90°$; but assume that θ_s is still large enough that fairly far from the hologram the object wave and secondary waves separate cleanly from each other. Then, taking account of the fact that the object wave field has the form $\mathcal{O}(x, y, z)e^{ikz}$, show that the secondary wave is the phase-conjugated object wave defined in this section, except that it is propagating in the $+z$ direction rather than $-z$ (i.e., it has been reflected through the $z = 0$ plane). Then use this and the discussion of phase conjugation in the text to show that the secondary wave carries an image that resides in front of the hologram and is turned inside out, as discussed near the end of Sec. 10.3. Show, further, that if θ_o is not $\ll 90°$ (but is $< 30°$, so θ_s is a real angle, and the secondary image actually exists), then the secondary image is changed by a distortion along the y direction. What is the nature of the distortion, a squashing or a stretch?

(b) Suppose that a hologram has been made with $\theta_o < 30°$. Show that it is possible to perform image reconstruction with a modified reference wave (different from Fig. 10.8) in such a manner that the secondary, phase-conjugated wave emerges precisely perpendicular to the hologram and undistorted.

10.5 Maxwell's Equations in a Nonlinear Medium; Nonlinear Dielectric Susceptibilities; Electro-Optic Effects

In nonlinear optics, one is often concerned with media that are electrically polarized with *polarization* (electric dipole moment per unit volume) **P** but have no free charges or currents and are unmagnetized. In such a medium the charge and current densities associated with the polarization are

$$\rho_P = -\mathbf{\nabla} \cdot \mathbf{P}, \qquad \mathbf{j}_P = \frac{\partial \mathbf{P}}{\partial t}, \qquad (10.16a)$$

and Maxwell's equations in SI units take the form

$$\mathbf{\nabla} \cdot \mathbf{E} = \frac{\rho_P}{\epsilon_0}, \quad \mathbf{\nabla} \cdot \mathbf{B} = 0, \quad \mathbf{\nabla} \times \mathbf{E} = -\frac{\partial \mathbf{B}}{\partial t}, \quad \mathbf{\nabla} \times \mathbf{B} = \mu_0 \left(\mathbf{j}_P + \epsilon_0 \frac{\partial \mathbf{E}}{\partial t} \right), \quad (10.16b)$$

which should be familiar. When rewritten in terms of the electric displacement vector

$$\mathbf{D} \equiv \epsilon_0 \mathbf{E} + \mathbf{P}, \qquad (10.17)$$

these Maxwell equations take the alternative form

$$\mathbf{\nabla} \cdot \mathbf{D} = 0, \qquad \mathbf{\nabla} \cdot \mathbf{B} = 0, \qquad \mathbf{\nabla} \times \mathbf{E} = -\frac{\partial \mathbf{B}}{\partial t}, \qquad \mathbf{\nabla} \times \mathbf{B} = \mu_0 \frac{\partial \mathbf{D}}{\partial t}, \quad (10.18)$$

which should also be familiar. By taking the curl of the third equation (10.16b), using the relation $\mathbf{\nabla} \times \mathbf{\nabla} \times \mathbf{E} = -\nabla^2 \mathbf{E} + \mathbf{\nabla}(\mathbf{\nabla} \cdot \mathbf{E})$, and combining with the time derivative of the fourth equation (10.16b) and with $\epsilon_0 \mu_0 = 1/c^2$ and $\mathbf{j}_P = \partial \mathbf{P}/\partial t$, we obtain the following wave equation for the electric field, sourced by the medium's polarization:

wave equation for light in a polarizable medium

$$\boxed{\nabla^2 \mathbf{E} - \mathbf{\nabla}(\mathbf{\nabla} \cdot \mathbf{E}) = \frac{1}{c^2} \frac{\partial^2 (\mathbf{E} + \mathbf{P}/\epsilon_0)}{\partial t^2}.} \qquad (10.19)$$

If the electric field is sufficiently weak and the medium is homogeneous and isotropic (the case treated in most textbooks on electromagnetic theory), the polarization **P** is proportional to the electric field: $\mathbf{P} = \epsilon_0 \chi_0 \mathbf{E}$, where χ_0 is the medium's electrical susceptibility. In this case the medium does not introduce any nonlinearities into Maxwell's equations, the right-hand side of Eq. (10.19) becomes $[(1 + \chi_0)/c^2]\partial^2 \mathbf{E}/\partial t^2$, the divergence of Eq. (10.19) implies that the divergence of **E** vanishes, and therefore Eq. (10.19) becomes the standard dispersionless wave equation:

$$\nabla^2 \mathbf{E} - \frac{\mathfrak{n}^2}{c^2} \frac{\partial^2 \mathbf{E}}{\partial t^2} = 0, \quad \text{with} \quad \mathfrak{n}^2 = 1 + \chi_0. \qquad (10.20)$$

In many dielectric media, however, a strong electric field can produce a polarization that is nonlinear in the field. In such nonlinear media, the general expression for the (real) polarization in terms of the (real) electric field is

polarization in a nonlinear medium

$$\boxed{P_i = \epsilon_0 (\chi_{ij} E_j + 2d_{ijk} E_j E_k + 4\chi_{ijkl} E_j E_k E_l + \ldots),} \qquad (10.21)$$

where we sum over repeated indices. Here χ_{ij}, the linear susceptibility, is proportional to the 3-dimensional metric, $\chi_{ij} = \chi_0 g_{ij} = \chi_0 \delta_{ij}$, if the medium is isotropic (i.e., if all directions in it are equivalent), but otherwise it is tensorial; and the d_{ijk} and χ_{ijkl} are nonlinear susceptibilities, referred to as *second-order* and *third-order*, respectively, because of the two and three **E** terms that multiply them in Eq. (10.21). The normalizations used for these second- and third-order susceptibilities differ from one researcher to another: sometimes the factor ϵ_0 is omitted in Eq. (10.21); occasionally the factors of 2 and 4 are omitted. A compressed 2-index notation is sometimes used for the components of d_{ijk}; see Box 10.2 in Sec. 10.6.

second-order and third-order nonlinear susceptibilities

With **P** given by Eq. (10.21), the wave equation (10.19) becomes

$$\nabla^2 \mathbf{E} - \nabla(\nabla \cdot \mathbf{E}) - \frac{1}{c^2}\boldsymbol{\epsilon} \cdot \frac{\partial^2 \mathbf{E}}{\partial t^2} = \frac{1}{c^2 \epsilon_0} \frac{\partial^2 \mathbf{P}^{\mathrm{NL}}}{\partial t^2}, \quad \text{where} \quad \epsilon_{ij} = \delta_{ij} + \chi_{ij}$$

dielectric tensor ϵ_{ij}

(10.22a)

is the "dielectric tensor," and $\mathbf{P}^{\mathrm{NL}}$ is the nonlinear part of the polarization:

$$P_i^{\mathrm{NL}} = \epsilon_0(2 d_{ijk} E_j E_k + 4\chi_{ijkl} E_j E_k E_l + \dots).$$

(10.22b)

When $\mathbf{P}^{\mathrm{NL}}$ is strong enough to be important and a monochromatic wave at frequency ω enters the medium, the nonlinearities lead to harmonic generation—the production of secondary waves with frequencies 2ω, 3ω, $\dots$; see Secs. 10.7.1 and 10.8.1. As a result, an electric field in the medium cannot oscillate at just one frequency, and each of the electric fields in expression (10.22b) for the nonlinear polarization must be a sum of pieces with different frequencies. Because *the susceptibilities can depend on frequency*, this means that, when using expression (10.21), one sometimes must break up P_i and each E_i into its frequency components and use different values of the susceptibility to couple the different frequencies together. For example, one of the terms in Eq. (10.22b) will become

harmonic generation by nonlinearities

$$P_i^{(4)} = 4\epsilon_0 \chi_{ijkl} E_j^{(1)} E_k^{(2)} E_l^{(3)},$$

(10.23)

where $E_j^{(n)}$ oscillates at frequency ω_n, $P_i^{(4)}$ oscillates at frequency ω_4, and χ_{ijkl} depends on the four frequencies $\omega_1, \dots, \omega_4$. Although this is complicated in the general case, in most practical applications, resonant coupling (or equivalently, energy and momentum conservation for photons) guarantees that only a single set of frequencies is important, and the resulting analysis simplifies substantially (see, e.g., Sec. 10.6.1).

Because all the tensor indices on the susceptibilities except the first index get contracted into the electric field in expression (10.21), we are free (and it is conventional) to define the susceptibilities as symmetric under interchange of any pair of indices that does not include the first. When [as has been tacitly assumed in Eq. (10.21)] there is no hysteresis in the medium's response to the electric field, the energy density of interaction between the polarization and the electric field is

polarizational energy density

$$U = \epsilon_0 \left(\frac{\chi_{ij} E_i E_j}{2} + \frac{2d_{ijk} E_i E_j E_k}{3} + \frac{4\chi_{ijkl} E_i E_j E_k E_l}{4} + \cdots \right), \quad (10.24a)$$

and the polarization is related to this energy of interaction, in Cartesian coordinates, by

$$P_i = \frac{\partial U}{\partial E_i}, \quad (10.24b)$$

which agrees with Eq. (10.21) *providing the susceptibilities are symmetric under interchange of all pairs of indices, including the first.* We shall assume such symmetry.[3] If the crystal is isotropic (as will be the case if it has cubic symmetry and reflection symmetry), then each of its tensorial susceptibilities is constructable from the metric $g_{ij} = \delta_{ij}$ and a single scalar susceptibility (see Ex. 10.7):[4]

susceptibilities for isotropic crystal

$$\chi_{ij} = \chi_0 g_{ij}, \quad d_{ijk} = 0,$$
$$\chi_{ijkl} = \tfrac{1}{3}\chi_4 (g_{ij} g_{kl} + g_{ik} g_{jl} + g_{il} g_{jk}), \quad \chi_{ijklm} = 0, \ldots \quad (10.25)$$

A simple model of a crystal that explains how nonlinear susceptibilities can arise is the following. Imagine each ion in the crystal as having a valence electron that can oscillate in response to a sinusoidal electric field. The electron can be regarded as residing in a potential well, which, for low-amplitude oscillations, is very nearly harmonic (potential energy quadratic in displacement; restoring force proportional to displacement; "spring constant" independent of displacement). However, if the electron's displacement from equilibrium becomes a significant fraction of the interionic distance, it will begin to feel the electrostatic attraction of the neighboring ions, and its spring constant will weaken. So the potential the electron sees is really not that of a harmonic oscillator, but rather that of an *anharmonic oscillator*, $V(x) = \alpha x^2 - \beta x^3 + \cdots$, where x is the electron's displacement from equilibrium. The nonlinearities in this potential cause the electron's amplitude of oscillation, when driven by a sinusoidal electric field,

---

3. For further details see, e.g., Sharma (2006, Sec. 14.3) or Yariv (1989, Secs. 16.2 and 16.3). In a lossy medium, symmetry on the first index is lost; see Yariv and Yeh (2007, Sec. 8.1).

4. There is a caveat to these symmetry arguments. When the nonlinear susceptibilities depend significantly on the frequencies of the three or four waves, then these simple symmetries can be broken. For example, the third-order susceptibility χ_{ijkl} for an isotropic medium depends on which of the three input waves is paired with the output wave in Eq. (10.25); so when one orders the input waves with wave 1 on the j index, 2 on the k index, and 3 on the l index (and output 4 on the i index), the three terms in χ_{ijkl} [Eq. (10.25)] have different scalar coefficients. We ignore this subtlety in the remainder of this chapter. (For details, see, e.g., Sharma, 2006, Sec. 14.3.)

to be nonlinear in the field strength, and that nonlinear displacement causes the crystal's polarization to be nonlinear (e.g., Yariv, 1989, Sec. 16.3). For most crystals, the spatial arrangement of the ions causes the electron's potential energy V to be different for displacements in different directions, which causes the nonlinear susceptibilities to be anisotropic.

Because the total energy required to liberate the electron from its lattice site is roughly 1 eV and the separation between lattice sites is $\sim 10^{-10}$ m, the characteristic electric field for strong instantaneous nonlinearities is ~ 1 V$(10^{-10}$ m$)^{-1} = 10^{10}$ V m$^{-1} = 1$ V $(100$ pm$)^{-1}$. Correspondingly, since d_{ijk} has dimensions 1/(electric field) and χ_{ijkl} has dimensions 1/(electric field)^2, rough upper limits on their Cartesian components are

$$d_{ijk} \lesssim 100 \text{ pm V}^{-1}, \qquad \chi_4 \sim \chi_{ijkl} \lesssim \left(100 \text{ pm V}^{-1}\right)^2. \qquad (10.26)$$

magnitudes of nonlinear
susceptibilities

For comparison, because stronger fields will pull electrons out of solids, the strongest continuous-wave electric fields that occur in practical applications are $E \sim 10^6$ V m^{-1}, corresponding to maximum intensities $F \sim 1$ kW mm$^{-2} = 1$ GW m^{-2}. These numbers dictate that, unless the second-order d_{ijk} are suppressed by isotropy, they will produce much larger effects than the third-order χ_{ijkl}, which in turn will dominate over all higher orders.

In the next few sections, we explore how the nonlinear susceptibilities produce nonlinear couplings of optical waves. There is, however, another application that we must mention in passing. When a slowly changing, non-wave electric field E_k is applied to a nonlinear medium, it can be thought of as producing a change in the linear dielectric tensor for waves $\Delta \chi_{ij} = 2(d_{ijk} + d_{ikj})E_k+$ quadratic terms [cf. Eq. (10.22b)]. This is an example (Boyd, 2008) of an *electro-optic effect*: the modification of optical properties of a medium by an applied electric field. Electro-optic effects are important in modern optical technology. For example, Pockels cells (used to modulate Gaussian light beams), optical switches (used in Q-switched lasers), and liquid-crystal displays (used for computer screens and television screens) are based on electro-optic effects. For some details of several important electro-optic effects and their applications, see, for example, Yariv and Yeh (2007, Chap. 9).

electro-optic effects

EXERCISES

Exercise 10.7 *Derivation and Example: Nonlinear Susceptibilities for an Isotropic Medium*
Explain why the nonlinear susceptibilities for an isotropic medium have the forms given in Eq. (10.25). [Hint: Use the facts that the χs must be symmetric in all their indices, and that, because the medium is isotropic, they must be constructable from

the only isotropic tensors available to us, the (symmetric) metric tensor g_{ij} and the (antisymmetric) Levi-Civita tensor ϵ_{ijk}.] What are the corresponding forms, in an isotropic medium, of χ_{ijklmn} and $\chi_{ijklmnp}$? [Note: We will encounter an argument similar to this, in Ex. 28.1, for the form of the Riemann tensor in an isotropic universe.]

10.6

10.6 Three-Wave Mixing in Nonlinear Crystals

10.6.1

10.6.1 Resonance Conditions for Three-Wave Mixing

When a beam of light is sent through a nonlinear crystal, the nonlinear suscepti- bilities produce wave-wave mixing. The mixing due to the second-order suscep- tibility d_{ijk} is called *three-wave mixing,* because three electric fields appear in the polarization-induced interaction energy, Eq. (10.24a). The mixing produced by the third-order χ_{ijkl} is similarly called *four-wave mixing.* Three-wave mixing dominates in an anisotropic medium, but it is suppressed when the medium is isotropic, leaving four-wave mixing as the leading-order nonlinearity.

three-wave and four-wave mixing

For use in our analyses of three-wave mixing, in Box 10.2 we list the second-order susceptibilities and some other properties of several specific nonlinear crystals.

Let us examine three-wave mixing in a general anisotropic crystal. Because the nonlinear susceptibilities are so small [i.e., because the input wave will generally be far weaker than 10^{10} V m^{-1} = 1 V(100 pm)$^{-1}$], the nonlinearities can be regarded as small perturbations. Suppose that two waves, labeled $n = 1$ and $n = 2$, are in- jected into the anisotropic crystal, and let their wave vectors be $\mathbf{k}_n$ when one ignores the (perturbative) nonlinear susceptibilities but keeps the large linear χ_{ij}. Because χ_{ij} is an anisotropic function of frequency, the dispersion relation $\Omega(\mathbf{k})$ for these waves (ignoring the nonlinearities) will typically be anisotropic. The frequencies of the two input waves satisfy the dispersion relation, $\omega_n = \Omega(\mathbf{k}_n)$, and the waves' forms are

$$E_j^{(n)} = \Re\left(\mathcal{A}_j^{(n)} e^{i(\mathbf{k}_n \cdot \mathbf{x} - \omega_n t)}\right) = \frac{1}{2}\left(\mathcal{A}_j^{(n)} e^{i(\mathbf{k}_n \cdot \mathbf{x} - \omega_n t)} + \mathcal{A}_j^{(n)*} e^{i(-\mathbf{k}_n \cdot \mathbf{x} + \omega_n t)}\right), \quad (10.27)$$

where we have denoted their vectorial complex amplitudes by $\mathcal{A}_j^{(n)}$. We adopt the convention that wave 1 is the wave with the larger frequency, so $\omega_1 - \omega_2 \geq 0$.

These two input waves couple, via the second-order nonlinear susceptibility d_{ijk}, to produce the following contribution to the medium's nonlinear polarization vector:

polarization for 3-wave mixing

$$P_i^{(3)} = 2\epsilon_0 d_{ijk}\, 2E_j^{(1)} E_k^{(2)}$$

$$= \epsilon_0 d_{ijk}\left(\mathcal{A}_j^{(1)} \mathcal{A}_k^{(2)} e^{i(\mathbf{k}_1 + \mathbf{k}_2)\cdot\mathbf{x}} e^{i(\omega_1 + \omega_2)t} + \mathcal{A}_j^{(1)} \mathcal{A}_k^{(2)*} e^{i(\mathbf{k}_1 - \mathbf{k}_2)\cdot\mathbf{x}} e^{i(\omega_1 - \omega_2)t} + \text{cc}\right),$$

$$(10.28)$$

BOX 10.2. PROPERTIES OF SOME ANISOTROPIC, NONLINEAR CRYSTALS

NOTATION FOR SECOND-ORDER SUSCEPTIBILITIES

In tabulations of the second-order nonlinear susceptibilities d_{ijk}, a compressed two-index notation d_{ab} is often used, with the indices running over

$$a: 1 = x, \quad 2 = y, \quad 3 = z,$$
$$b: 1 = xx, \quad 2 = yy, \quad 3 = zz, \quad 4 = yz = zy, \quad 5 = xz = zx, \quad 6 = xy = yx. \tag{1}$$

CRYSTALS WITH LARGE SECOND-ORDER SUSCEPTIBILITIES

The following crystals have especially large second-order susceptibilities:

$$\text{Te: tellurium} \quad d_{11} = d_{xxx} = 650 \text{ pm } \text{V}^{-1}$$
$$\text{CdGeAs}_2 \quad d_{36} = d_{zyx} = 450 \text{ pm } \text{V}^{-1}$$
$$\text{Se: selenium} \quad d_{11} = d_{xxx} = 160 \text{ pm } \text{V}^{-1}. \tag{2}$$

However, they are not widely used in nonlinear optics, because some of their other properties are unfavorable. By contrast, glasses containing tellurium or selenium have moderately large nonlinearities and are useful.

KH_2PO_4

Potassium dihydrogen phosphate (KDP) is among the most widely used nonlinear crystals in 2016, not because of its nonlinear susceptibilities (which are quite modest) but because (i) it can sustain large electric fields without suffering damage, (ii) it is highly birefringent (different light speeds in different directions and for different polarizations, which as we shall see in Sec. 10.6.3 is useful for phase matching), and (iii) it has large electro-optic coefficients (end of Sec. 10.5). At linear order, it is axisymmetric around the z-axis, and its indices of refraction and susceptibilities have the following dependence on wavelength λ (measured in microns), which we use in Sec. 10.6.3 and Fig. 10.12a:

$$(n_o)^2 = 1 + \chi_{xx} = 1 + \chi_{yy} = 2.259276 + \frac{0.01008956}{\lambda^2 - 0.012942625} + \frac{13.005522\lambda^2}{\lambda^2 - 400},$$

$$(n_e)^2 = 1 + \chi_{zz} = 2.132668 + \frac{0.008637494}{\lambda^2 - 0.012281043} + \frac{3.2279924\lambda^2}{\lambda^2 - 400}. \tag{3}$$

The second-order nonlinearities break the axisymmetry of KDP, giving rise to

$$d_{36} = d_{zyx} = 0.44 \text{ pm } \text{V}^{-1}. \tag{4}$$

Although this is three orders of magnitude smaller than the largest nonlinearities available, its smallness is compensated for by its ability to sustain large electric fields.

(continued)

BOX 10.2.   (continued)

KTiOPO$_4$

Potassium titanyl phosphate (KTP) is quite widely used in 2016 (e.g., in green laser pointers; Ex. 10.13). At linear order it is nonaxisymmetric, but with only modest birefringence: its indices of refraction along its three principal axes, at the indicated wavelengths, are

$$1{,}064 \text{ nm: } n_x = \sqrt{1+\chi_{xx}} = 1.740, \quad n_y = \sqrt{1+\chi_{yy}} = 1.747,$$

$$n_z = \sqrt{1+\chi_{zz}} = 1.830;$$

$$532 \text{ nm: } n_x = \sqrt{1+\chi_{xx}} = 1.779, \quad n_y = \sqrt{1+\chi_{yy}} = 1.790,$$

$$n_z = \sqrt{1+\chi_{zz}} = 1.887. \tag{5}$$

Its third-order nonlinearities are moderately large. In units of pm V^{-1}, they are

$$d_{31} = d_{zxx} = 6.5, \qquad d_{32} = d_{zyy} = 5.0, \qquad d_{33} = d_{zzz} = 13.7,$$

$$d_{24} = d_{xyz} = d_{xzy} = 7.6, \qquad d_{15} = d_{xxz} = d_{xzx} = 6.1. \tag{6}$$

Notice that symmetry on the first index is modestly broken: $d_{zxx} = 6.5 \neq d_{xxz} = 7.6$. This symmetry breaking is caused by the crystal's dissipating a small portion of the light power that drives it.

EVOLUTION OF MATERIALS

Over the past three decades materials scientists have found and developed nonlinear crystals with ever-improving properties. By the time you read this book, the most widely used crystals are likely to have changed.

where "cc" means complex conjugate.[5] This sinusoidally oscillating polarization produces source terms in Maxwell's equations (10.16b) and the wave equation (10.19): an oscillating, polarization-induced charge density $\rho_P = -\nabla \cdot \mathbf{P}^{(3)}$ and current density $\mathbf{j}_P = \partial \mathbf{P}^{(3)}/\partial t$. This polarization charge and current, like $\mathbf{P}^{(3)}$ itself [Eq. (10.28)], consist of two traveling waves, one with frequency and wave vector

resonance conditions and dispersion relation for new, third wave

$$\boxed{\omega_3 = \omega_1 + \omega_2, \qquad \mathbf{k}_3 = \mathbf{k}_1 + \mathbf{k}_2;} \tag{10.29a}$$

the other with frequency and wave vector

$$\boxed{\omega_3 = \omega_1 - \omega_2, \qquad \mathbf{k}_3 = \mathbf{k}_1 - \mathbf{k}_2.} \tag{10.29b}$$

---

5. The reason for the factor 2 in the definition $P_i = 2\epsilon_0 d_{ijk} E_j E_k$ is to guarantee a factor unity in Eq. (10.28) and in the resulting coupling constant κ of Eq. (10.38).

If either of these (ω_3, $\mathbf{k}_3$) satisfies the medium's dispersion relation $\omega = \Omega(\mathbf{k})$, then the polarization will generate an electromagnetic wave $E_j^{(3)}$ that propagates along in resonance with its polarization-vector source in the wave equation

$$\nabla^2 \mathbf{E}^{(3)} - \nabla(\nabla \cdot \mathbf{E}^{(3)}) + \frac{\omega_3^2}{c^2}\boldsymbol{\epsilon} \cdot \mathbf{E}^{(3)} = \frac{1}{c^2\epsilon_0}\frac{\partial^2 \mathbf{P}^{(3)}}{\partial t^2} \qquad (10.30)$$

[the frequency-ω_3 part of Eq. (10.22a)]. Therefore, this new electromagnetic wave, with frequency ω_3 and wave vector $\mathbf{k}_3$, will grow as it propagates.

For most choices of the input waves—most choices of $\{\mathbf{k}_1, \omega_1 = \Omega(\mathbf{k}_1), \mathbf{k}_2, \omega_2 = \Omega(\mathbf{k}_2)\}$—neither of the polarizations $\mathbf{P}^{(3)}$ will have a frequency $\omega_3 = \omega_1 \pm \omega_2$ and wave vector $\mathbf{k}_3 = \mathbf{k}_1 \pm \mathbf{k}_2$ that satisfy the medium's dispersion relation, and thus neither will be able to create a third electromagnetic wave resonantly; the wave-wave mixing is ineffective. However, for certain special choices of the input waves, resonant coupling will be achieved, and a strong third wave will be produced. See Sec. 10.6.3 for details.

In nonlinear optics, enforcing the resonance conditions (10.29), with all three waves satisfying their dispersion relations, is called *phase matching,* because it guarantees that the new wave propagates along in phase with the polarization produced by the two old waves.

phase matching

The resonance conditions (10.29) have simple quantum mechanical interpretations—a fact that is not at all accidental: quantum mechanics underlies the classical theory that we are developing. Each classical wave is carried by photons that have discrete energies $\mathcal{E}_n = \hbar\omega_n$ and discrete momenta $\mathbf{p}_n = \hbar\mathbf{k}_n$. The input waves are able to produce resonantly waves with $\omega_3 = \omega_1 \pm \omega_2$ and $\mathbf{k}_3 = \mathbf{k}_1 \pm \mathbf{k}_2$, if those waves satisfy the dispersion relation. Restated in quantum mechanical terms, the condition of resonance with the "+" sign rather than the "−" is

quantum description of resonance conditions

one photon created from two

$$\mathcal{E}_3 = \mathcal{E}_1 + \mathcal{E}_2, \qquad \mathbf{p}_3 = \mathbf{p}_1 + \mathbf{p}_2. \qquad (10.31a)$$

This has the quantum mechanical meaning that one photon of energy $\mathcal{E}_1$ and momentum $\mathbf{p}_1$, and another of energy $\mathcal{E}_2$ and momentum $\mathbf{p}_2$ combine together, via the medium's nonlinearities, and are annihilated (in the language of quantum field theory). By their annihilation they create a new photon with energy $\mathcal{E}_3 = \mathcal{E}_1 + \mathcal{E}_2$ and momentum $\mathbf{p}_3 = \mathbf{p}_1 + \mathbf{p}_2$. Thus the classical condition of resonance is the quantum mechanical condition of energy-momentum conservation for the sets of photons involved in a process of quantum annihilation and creation. For this process to proceed, not only must energy-momentum conservation be satisfied, but also all three photons must have energies and momenta that obey the photons' semiclassical hamiltonian relation $\mathcal{E} = H(\mathbf{p})$ (i.e., the dispersion relation $\omega = \Omega(\mathbf{k})$ with $H = \hbar\Omega$, $\mathcal{E} = \hbar\omega$, and $\mathbf{p} = \hbar\mathbf{k}$).

Similarly, the classical conditions of resonance with the "−" sign rather than the "+" can be written (after bringing photon 2 to the left-hand side) as

$$\mathcal{E}_3 + \mathcal{E}_2 = \mathcal{E}_1, \qquad \mathbf{p}_3 + \mathbf{p}_2 = \mathbf{p}_1. \tag{10.31b}$$

This has the quantum mechanical meaning that one photon of energy $\mathcal{E}_1$ and momentum $\mathbf{p}_1$ is annihilated, via the medium's nonlinearities, and from its energy and momentum two photons are created, with energies $\mathcal{E}_2$, $\mathcal{E}_3$ and momenta $\mathbf{p}_2$, $\mathbf{p}_3$ that satisfy energy-momentum conservation.

Resonance conditions play a major role in other areas of physics, whenever one deals with nonlinear wave-wave coupling or wave-particle coupling. In this book we meet them again in both classical language and quantum language when studying excitations of plasmas (Chap. 23).

10.6.2

10.6.2 Three-Wave-Mixing Evolution Equations in a Medium That Is Dispersion-Free and Isotropic at Linear Order

Consider the simple, idealized case where the linear part of the susceptibility χ_{jk} is isotropic and frequency independent, $\chi_{jk} = \chi_0 g_{jk}$; correspondingly, Maxwell's equations imply $\nabla \cdot \mathbf{E} = 0$. The Track-One part of this chapter will be confined to this idealized case. In Sec. 10.6.3 (Track Two), we treat the more realistic case, which has dispersion and anisotropy at linear order.

In our idealized case the dispersion relation, ignoring the nonlinearities, takes the simple, nondispersive form [which follows from Eq. (10.20)]:

$$\omega = \frac{c}{\mathfrak{n}} k, \quad \text{where} \quad k = |\mathbf{k}|, \quad \mathfrak{n} = \sqrt{1 + \chi_0}. \tag{10.32}$$

Consider three-wave mixing for waves 1, 2, and 3 that all propagate in the same z direction with wave numbers that satisfy the resonance condition $k_3 = k_1 + k_2$. The dispersion-free dispersion relation (10.32) guarantees that the frequencies will also resonate: $\omega_3 = \omega_1 + \omega_2$. If we write the new wave as $E_i^{(3)} = \Re(\mathcal{A}_i^{(3)} e^{i(k_3 z - \omega_3 t)}) = \frac{1}{2}\mathcal{A}_i^{(3)} e^{i(k_3 z - \omega_3 t)} + \text{cc}$, then its evolution equation (10.30), when combined with Eqs. (10.27) and (10.28), takes the form

$$\nabla^2 \left(\mathcal{A}_i^{(3)} e^{i(k_3 z - \omega_3 t)} \right) + \frac{\mathfrak{n}^2 \omega_3^2}{c^2} \mathcal{A}_i^{(3)} e^{i(k_3 z - \omega_3 t)} = -2 \frac{\omega_3^2}{c^2} d_{ijk} \mathcal{A}_j^{(1)} \mathcal{A}_k^{(2)} e^{i(k_3 z - \omega_3 t)}. \tag{10.33}$$

Using the dispersion relation (10.32) and the fact that the lengthscale on which wave 3 changes is long compared to its wavelength (which is always the case, because the fields are always much weaker than 10^{10} V m^{-1}), the left-hand side becomes $2ik_3 d\mathcal{A}_i^{(3)}/dz$, and Eq. (10.33) then becomes (with the aid of the dispersion relation) $d\mathcal{A}_i^{(3)}/dz = i(k_3/\mathfrak{n}^2)d_{ijk}\mathcal{A}_j^{(1)}\mathcal{A}_k^{(2)}$. This and similar computations for evolution of the other two waves (Ex. 10.8) give the following equations for the rates of change of the three waves' complex amplitudes:

$$\frac{d\mathcal{A}_i^{(3)}}{dz} = i\frac{k_3}{n^2}d_{ijk}\mathcal{A}_j^{(1)}\mathcal{A}_k^{(2)} \quad \text{at } \omega_3 = \omega_1 + \omega_2, \ k_3 = k_1 + k_2; \tag{10.34a}$$

$$\frac{d\mathcal{A}_i^{(1)}}{dz} = i\frac{k_1}{n^2}d_{ijk}\mathcal{A}_j^{(3)}\mathcal{A}_k^{(2)*} \quad \text{at } \omega_1 = \omega_3 - \omega_2, \ k_1 = k_3 - k_2; \tag{10.34b}$$

$$\frac{d\mathcal{A}_i^{(2)}}{dz} = i\frac{k_2}{n^2}d_{ijk}\mathcal{A}_j^{(3)}\mathcal{A}_k^{(1)*} \quad \text{at } \omega_2 = \omega_3 - \omega_1, \ k_2 = k_3 - k_1. \tag{10.34c}$$

Therefore, each wave's amplitude changes with distance z traveled at a rate proportional to the product of the field strengths of the other two waves.

It is instructive to rewrite the evolution equations (10.34) in terms of *renormalized scalar amplitudes* $\mathfrak{A}_n$ and *unit-normed polarization vectors* $f_j^{(n)}$ for the three waves $n = 1, 2, 3$:

renormalized wave amplitudes

$$\mathcal{A}_j^{(n)} = \sqrt{\frac{2k_n}{\epsilon_0 n^2}}\, \mathfrak{A}_n\, f_j^{(n)} = \sqrt{\frac{2\omega_n}{\epsilon_0 c\, n}}\, \mathfrak{A}_n\, f_j^{(n)}. \tag{10.35}$$

This renormalization is motivated by the fact that $|\mathfrak{A}_n|^2$ is proportional to the flux of quanta $dN_n/dA\,dt$ associated with wave n. Specifically, the energy density in wave n is (neglecting nonlinearities) $U = \epsilon_o(1 + \chi_o)\overline{E^2} = \frac{1}{2}\epsilon_o n^2 |\mathcal{A}^{(n)}|^2$ (where the bar means time average); the energy flux is this U times the wave speed c/n:

$$\boxed{F_n = \frac{1}{2}\epsilon_o nc |\mathcal{A}^{(n)}|^2 = \omega_n |\mathfrak{A}_n|^2;} \tag{10.36}$$

and the flux of quanta is this F_n divided by the energy $\mathcal{E}_n = \hbar\omega_n$ of each quantum: $dN_n/dA\,dt = |\mathfrak{A}_n|^2/\hbar$, where dA is a unit area orthogonal to $\mathbf{k}_n$.

The three-wave-mixing evolution equations (10.34), rewritten in terms of the renormalized amplitudes, take the simple form

three-wave mixing evolution equations in an isotropic, dispersion-free medium

$$\frac{d\mathfrak{A}_3}{dz} = i\kappa\, \mathfrak{A}_1\mathfrak{A}_2, \quad \frac{d\mathfrak{A}_1}{dz} = i\kappa\, \mathfrak{A}_3\mathfrak{A}_2^*, \quad \frac{d\mathfrak{A}_2}{dz} = i\kappa\, \mathfrak{A}_3\mathfrak{A}_1^*;$$

$$\kappa = \sqrt{\frac{2\omega_1\omega_2\omega_3}{\epsilon_0 c^3 n^3}}\, d_{ijk}\, f_i^{(1)} f_j^{(2)} f_k^{(3)}. \tag{10.37}$$

It is straightforward to verify that these evolution equations guarantee energy conservation $d(F_1 + F_2 + F_3)/dz = 0$, with F_n given by Eq. (10.36). Therefore, at least one wave will grow and at least one wave will decay due to three-wave mixing.

When waves 1 and 2 are the same wave, the three-wave mixing leads to frequency doubling: $\omega_3 = 2\omega_1$. In this case, the nonlinear polarization that produces the third wave is $P_i = \epsilon_0 d_{ijk}E_j^{(1)}E_k^{(1)}$, by contrast with that when waves 1 and 2 are different, $P_i = 2\epsilon_0 d_{ijk}E_j^{(1)}E_j^{(2)}$ [Eq. (10.28)]. [In the latter case the factor 2 arises because we are dealing with cross terms in $(E_j^{(1)} + E_j^{(2)})(E_k^{(1)} + E_k^{(2)})$.] Losing the factor 2 and

frequency doubling

making wave 2 the same as wave 1 leads to an obvious modification of the evolution equations (10.37):

evolution equations for
frequency doubling

$$\frac{d\mathfrak{A}_3}{dz} = \frac{i\kappa}{2}(\mathfrak{A}_1)^2, \quad \frac{d\mathfrak{A}_1}{dz} = i\kappa\mathfrak{A}_3\mathfrak{A}_1^*; \quad \kappa = \sqrt{\frac{2\omega_1^2\omega_3}{\epsilon_0 c^3 n^3}}\, d_{ijk}\, f_i^{(1)} f_j^{(1)} f_k^{(3)}. \quad (10.38)$$

Once again, it is easy to verify energy conservation, $d(F_1 + F_3)/dz = 0$. We discuss frequency doubling in Sec. 10.7.1.

EXERCISES

Exercise 10.8 *Derivation: Evolution Equations in Idealized Three-Wave Mixing*
Derive Eqs. (10.34b) and (10.34c) for the amplitudes of waves 1 and 2 produced by three-wave mixing.

10.6.3

10.6.3 Three-Wave Mixing in a Birefringent Crystal: Phase Matching and Evolution Equations [T2]

ORDINARY WAVES, EXTRAORDINARY WAVES, AND DISPERSION RELATIONS

In reality, all nonlinear media have frequency-dependent dispersion relations and

birefringent crystal

most are anisotropic at linear order and therefore birefringent (different wave speeds in different directions). An example is the crystal KDP (Box 10.2), which is symmetric around the z-axis and has indices of refraction[6]

$$n_o = \sqrt{1 + \chi_{xx}} = \sqrt{1 + \chi_{yy}}, \quad n_e = \sqrt{1 + \chi_{zz}} \quad (10.39)$$

that depend on the light's wave number $k = 2\pi/\lambda$ in the manner shown in Fig. 10.12a and in Eq. (3) of Box 10.2. The subscript o stands for *ordinary*; e, for *extraordinary*; see the next paragraph.

Maxwell's equations imply that, in this crystal, for plane, monochromatic waves propagating in the x-z plane at an angle θ to the symmetry axis [$\mathbf{k} = k(\sin\theta\mathbf{e}_x + \cos\theta\mathbf{e}_z)$], there are two dispersion relations corresponding to the two polarizations of the electric field:

1. If $\mathbf{E}$ is orthogonal to the symmetry axis, then (as is shown in Ex. 10.9), it must also be orthogonal to the propagation direction (i.e., must point in the $\mathbf{e}_y$ direction), and the dispersion relation is

$$\frac{\omega/k}{c} = \text{(phase speed in units of speed of light)} = \frac{1}{n_o}, \quad (10.40a)$$

---

6. For each wave, the index of refraction is the ratio of light's vacuum speed c to the wave's phase speed $V_{\text{ph}} = \omega/k$.

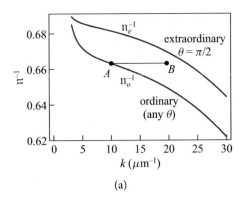
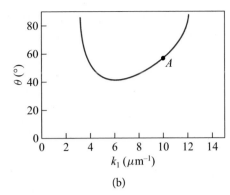

(a)   (b)

FIGURE 10.12 (a) The inverse of the index of refraction n^{-1} (equal to the phase speed in units of the speed of light) for electromagnetic waves propagating at an angle θ to the symmetry axis of a KDP crystal, as a function of wave number k in reciprocal microns. See Eq. (10.40a) for the lower curve and Eq. (10.40b) with $\theta = \pi/2$ for the upper curve. For extraordinary waves propagating at an arbitrary angle θ to the crystal's symmetry axis, n^{-1} is a mean [Eq. (10.40b)] of the two plotted curves. The plotted curves are fit by the analytical formulas (3) of Box 10.2. (b) The angle θ to the symmetry axis at which ordinary waves with wave number k_1 (e.g., point A) must propagate for three-wave mixing to be able to produce frequency-doubled or phase-conjugated extraordinary waves (e.g., point B).

independent of the angle θ. These waves are called *ordinary*, and their phase speed (10.40a) is the lower curve in Fig. 10.12a; at $k = 10\ \mu\text{m}^{-1}$ (point A), the phase speed is $0.663c$, while at $k = 20\ \mu\text{m}^{-1}$, it is $0.649c$.

2. If **E** is not orthogonal to the symmetry axis, then (Ex. 10.9) it must lie in the plane formed by **k** and the symmetry axis (the x-z) plane, with $E_x/E_z = -(n_e/n_o)^2 \cot\theta$ (which means that **E** is not orthogonal to the propagation direction unless the crystal is isotropic, $n_e = n_o$, which it is not); and the dispersion relation is

$$\frac{\omega/k}{c} = \frac{1}{n} = \sqrt{\frac{\cos^2\theta}{n_o^2} + \frac{\sin^2\theta}{n_e^2}}. \tag{10.40b}$$

In this case the waves are called *extraordinary*.[7] As the propagation direction varies from parallel to the symmetry axis ($\cos\theta = 1$) to perpendicular ($\sin\theta = 1$), this extraordinary phase speed varies from c/n_o (the lower curve in Fig. 10.12; $0.663c$ at $k = 10\ \mu\text{m}^{-1}$), to c/n_e (the upper curve; $0.681c$ at $k = 10\ \mu\text{m}^{-1}$).

---

7. When studying perturbations of a cold, magnetized plasma (Sec. 21.5.3) we will meet two wave modes that have these same names: ordinary and extraordinary. However, because of the physical differences between an axially symmetric magnetized plasma and an axially symmetric birefringent crystal, the physics of those plasma modes (e.g., the direction of their oscillating electric field) is rather different from that of the crystal modes studied here.

This birefringence enables one to achieve phase matching (satisfy the resonance conditions) in three-wave mixing. As an example, consider the resonance conditions for a *frequency-doubling device* (discussed in further detail in Sec. 10.7.1): one in which the two input waves are identical, so $\mathbf{k}_1 = \mathbf{k}_2$ and $\mathbf{k}_3 = 2\mathbf{k}_1$ point in the same direction. Let this common propagation direction be at an angle θ to the symmetry axis. Then the resonance conditions reduce to the demands that the output wave number be twice the input wave number, $k_3 = 2k_1$, and the output phase speed be the same as the input phase speed, $\omega_3/k_3 = \omega_1/k_1$. Now, for waves of the same type (both ordinary or both extraordinary), the phase speed is a monotonic decreasing function of wave number [Fig. 10.12a and Eqs. (10.40)], so there is no choice of propagation angle θ that enables these resonance conditions to be satisfied. The only way to satisfy them is by using ordinary input waves and extraordinary output waves, and then only for a special, frequency-dependent propagation direction. This technique is called *type I phase matching*; "type I" because there are other techniques for phase matching (i.e., for arranging that the resonance conditions be satisfied; see, e.g., Table 8.4 of Yariv and Yeh, 2007).

As an example, if the input waves are ordinary, with $k_1 = 10\ \mu\mathrm{m}^{-1}$ (approximately the value for light from a ruby laser; point A in Fig. 10.12a), then the output waves must be extraordinary and must have the same phase speed as the input waves (same height in Fig. 10.12a) and have $k_3 = 2k_1 = 20\ \mu\mathrm{m}^{-1}$ (i.e., point B). This phase speed is between $c/\mathrm{n}_e(2k_1)$ and $c/\mathrm{n}_o(2k_1)$, and thus can be achieved for a special choice of propagation angle: $\theta = 56.7°$ (point A in Fig. 10.12b). In general, Eqs. (10.40) imply that the unique propagation direction θ at which the resonance conditions can be satisfied is the following function of the input wave number k_1:

$$\sin^2\theta = \frac{1/\mathrm{n}_o^2(k_1) - 1/\mathrm{n}_o^2(2k_1)}{1/\mathrm{n}_e^2(2k_1) - 1/\mathrm{n}_o^2(2k_1)}. \tag{10.41}$$

This resonance angle is plotted as a function of wave number for KDP in Fig. 10.12b.

This special case of identical input waves illustrates a general phenomenon: *at fixed input frequencies, the resonance conditions can be satisfied only for special, discrete input and output directions.*

For our frequency-doubling example, the extraordinary dispersion relation (10.40b) for the output wave can be rewritten as

$$\omega = \frac{ck}{\mathrm{n}} = \Omega_e(\mathbf{k}) = c\sqrt{\frac{k_z^2}{\mathrm{n}_o(k)^2} + \frac{k_x^2}{\mathrm{n}_e(k)^2}}, \quad \text{where} \quad k = \sqrt{k_x^2 + k_z^2}. \tag{10.42}$$

Correspondingly, the group velocity[8] $V_g^j = \partial\Omega/\partial k_j$ for the output wave has components

$$V_g^x = V_{\rm ph}\sin\theta\left(\frac{n^2}{n_e^2} - \frac{n^2\cos^2\theta}{n_o^2}\frac{d\ln n_o}{d\ln k} - \frac{n^2\sin^2\theta}{n_e^2}\frac{d\ln n_e}{d\ln k}\right),$$

$$V_g^z = V_{\rm ph}\cos\theta\left(\frac{n^2}{n_o^2} - \frac{n^2\cos^2\theta}{n_o^2}\frac{d\ln n_o}{d\ln k} - \frac{n^2\sin^2\theta}{n_e^2}\frac{d\ln n_e}{d\ln k}\right), \qquad (10.43)$$

where $V_{\rm ph} = \omega/k = c/n$ is the phase velocity. For an ordinary input wave with $k_1 = 10\ \mu{\rm m}^{-1}$ (point A in Fig. 10.12) and an extraordinary output wave with $k_3 = 20\ \mu{\rm m}^{-1}$ (point B), these formulas give for the direction of the output group velocity (direction along which the output waves grow) $\theta_g = \arctan(V_g^x/V_g^z) = 58.4^\circ$, compared to the direction of the common input-output phase velocity $\theta = 56.7^\circ$. They give for the magnitude of the group velocity $V_g = 0.628c$, compared to the common phase velocity $v_{\rm ph} = 0.663c$. Thus, the differences between the group velocity and the phase velocity are small, but they do differ.

EVOLUTION EQUATIONS

Once one has found wave vectors and frequencies that satisfy the resonance conditions, the evolution equations for the two (or three) coupled waves have the same form as in the idealized dispersion-free, isotropic case [Eqs. (10.38) or (10.37)], but with the following minor modifications.

Let planar input waves impinge on a homogeneous, nonlinear crystal at some plane $z = 0$ and therefore (by symmetry) have energy fluxes inside the crystal that evolve as functions of z only: $\mathbf{F}_n = \mathbf{F}_n(z)$ for waves $n = 1$ and 3 in the case of frequency doubling (or 1, 2, and 3 in the case of three different waves). Then energy conservation dictates that

$$\frac{d}{dz}\sum_n F_{nz} = 0, \qquad (10.44)$$

where $F_{nz}(z)$ is the z component of the energy flux for wave n. It is convenient to define a complex amplitude $\mathfrak{A}_n$ for wave n that is related to the wave's complex electric field amplitude by an analog of Eq. (10.35):

$$\mathcal{A}_j^{(n)} = \zeta_n\sqrt{\frac{2\omega_n}{\epsilon_0 c\, n_n}}\,\mathfrak{A}_n f_j^{(n)}. \qquad (10.45)$$

renormalized amplitude in birefringent crystal

Here $f_j^{(n)}$ is the wave's polarization vector [Eq. (10.35)], n_n is its index of refraction (defined by $\omega_n/k_n = c/n_n$), and ζ_n is some positive real constant that depends on the

---

8. Since a wave's energy travels with the group velocity, it must be that $\mathbf{V}_g = \mathbf{E}\times\mathbf{H}/U$, where U is the wave's energy density, $\mathbf{E}\times\mathbf{H}$ is its Poynting vector (energy flux), and $\mathbf{H} = \mathbf{B}/\mu_0$ (in our dielectric medium). It can be shown explicitly that, indeed, this is the case.

relative directions of $\mathbf{k}_n$, $\mathbf{f}^{(n)}$, and $\mathbf{e}_z$ and has a value ensuring that

$$F_{n\,z} = \omega_n |\mathfrak{A}_n|^2 \qquad (10.46)$$

[same as Eq. (10.36) but with F_n replaced by $F_{n\,z}$]. Since the energy flux is $\hbar\omega_n$ times the photon-number flux, this equation tells us that $|\mathfrak{A}_n|^2/\hbar$ is the photon-number flux (just like the idealized case).

Because the evolution equations involve the same photon creation and annihilation processes as in the idealized case, they must have the same mathematical form as in that case [Eqs. (10.38) or (10.37)], except for the magnitude of the coupling constant. (For a proof, see Ex. 10.10.) Specifically, for frequency doubling of a wave 1 to produce wave 3 ($\omega_3 = 2\omega_1$), the resonant evolution equations and coupling constant are

three-wave mixing evolution equations in a birefringent crystal

$$\frac{d\mathfrak{A}_3}{dz} = \frac{i\kappa}{2}(\mathfrak{A}_1)^2, \quad \frac{d\mathfrak{A}_1}{dz} = i\kappa\,\mathfrak{A}_3\mathfrak{A}_1^*; \quad \kappa = \beta\sqrt{\frac{2\omega_1^2\omega_3}{\epsilon_0 c^3 n_1^2 n_3}}\,d_{ijk}\,f_i^{(1)}f_j^{(1)}f_k^{(3)}$$

$$(10.47)$$

[cf. Eqs. (10.38) for the idealized case]. For resonant mixing of three different waves ($\omega_3 = \omega_1 + \omega_2$), they are

$$\frac{d\mathfrak{A}_3}{dz} = i\kappa\,\mathfrak{A}_1\mathfrak{A}_2, \quad \frac{d\mathfrak{A}_1}{dz} = i\kappa\,\mathfrak{A}_3\mathfrak{A}_2^*, \quad \frac{d\mathfrak{A}_2}{dz} = i\kappa\,\mathfrak{A}_3\mathfrak{A}_1^*;$$

$$\kappa = \beta'\sqrt{\frac{2\omega_1\omega_2\omega_3}{\epsilon_0 c^3 n_1 n_2 n_3}}\,d_{ijk}\,f_i^{(1)}f_j^{(2)}f_k^{(3)}$$

$$(10.48)$$

[cf. Eqs. (10.37) for the idealized case]. Here β and β' are constants of order unity that depend on the relative directions of $\mathbf{e}_z$ and the wave vectors $\mathbf{k}_n$ and polarization vectors $\mathbf{f}^{(n)}$; see Ex. 10.10.

It is useful to keep in mind the following magnitudes of the quantities that appear in these three-wave-mixing equations (Ex. 10.11):

$$F_n \lesssim 1\,\mathrm{GW\,m^{-2}}, \quad |\mathfrak{A}_n| \lesssim 10^{-3}\,\mathrm{J^{1/2}\,m^{-1}}, \quad \kappa \lesssim 10^5\,\mathrm{J^{-1/2}}, \quad |\kappa\,\mathfrak{A}_n| \lesssim 1\,\mathrm{cm^{-1}}.$$

$$(10.49)$$

We use the evolution equations (10.47) and (10.48) in Sec. 10.7 to explore several applications of three-wave mixing.

One can reformulate the equations of three-wave mixing in fully quantum mechanical language, with a focus on the mean occupation numbers of the wave modes. This is commonly done in plasma physics; in Sec. 23.3.6 we discuss the example of coupled electrostatic waves in a plasma.

Exercise 10.9 **Example: Dispersion Relation for an Anisotropic Medium* [T2]
Consider a wave propagating through a dielectric medium that is anisotropic, but not necessarily—for the moment—axisymmetric. Let the wave be sufficiently weak that nonlinear effects are unimportant. Then the nonlinear wave equation (10.22a) takes the linear form

$$-\nabla^2 \mathbf{E} + \nabla(\nabla \cdot \mathbf{E}) = -\frac{1}{c^2}\boldsymbol{\epsilon} \cdot \frac{\partial^2 \mathbf{E}}{\partial t^2}. \tag{10.50}$$

(a) Specialize to a monochromatic plane wave with angular frequency ω and wave vector $\mathbf{k}$. Show that the wave equation (10.50) reduces to

$$L_{ij}E_j = 0, \quad \text{where } L_{ij} = k_i k_j - k^2 \delta_{ij} + \frac{\omega^2}{c^2}\epsilon_{ij}. \tag{10.51a}$$

This equation says that E_j is an eigenvector of L_{ij} with vanishing eigenvalue, which is possible if and only if

$$\det \|L_{ij}\| = 0. \tag{10.51b}$$

This vanishing determinant is the waves' dispersion relation. We use it in Chap. 21 to study waves in plasmas.

(b) Next specialize to an axisymmetric medium, and orient the symmetry axis along the z direction, so the only nonvanishing components of the dielectric tensor ϵ_{ij} are $\epsilon_{11} = \epsilon_{22}$ and ϵ_{33}. Let the wave propagate in a direction $\hat{\mathbf{k}}$ that makes an angle θ to the symmetry axis. Show that in this case L_{ij} has the form

$$\|L_{ij}\| = k^2 \begin{Vmatrix} (n_o/n)^2 - \cos^2\theta & 0 & \sin\theta\cos\theta \\ 0 & (n_o/n)^2 - 1 & 0 \\ \sin\theta\cos\theta & 0 & (n_e/n)^2 - \sin^2\theta \end{Vmatrix}, \tag{10.52a}$$

and the dispersion relation (10.51b) reduces to

$$\left(\frac{1}{n^2} - \frac{1}{n_o^2}\right)\left(\frac{1}{n^2} - \frac{\cos^2\theta}{n_o^2} - \frac{\sin^2\theta}{n_e^2}\right) = 0, \tag{10.52b}$$

where $1/n = \omega/kc$, $n_o = \sqrt{\epsilon_{11}} = \sqrt{\epsilon_{22}}$, and $n_e = \sqrt{\epsilon_{33}}$, in accord with Eq. (10.39).

(c) Show that this dispersion relation has the two solutions (ordinary and extraordinary) discussed in the text, Eqs. (10.40a) and (10.40b), and show that the electric fields associated with these two solutions point in the directions described in the text.

Exercise 10.10 **Derivation and Example: Evolution Equations for Realistic Wave-Wave Mixing* [T2]
Derive the evolution equations (10.48) for three-wave mixing. [The derivation of those (10.47) for frequency doubling is similar.] You could proceed as follows.

(a) Insert expressions (10.27) and (10.28) into the general wave equation (10.30) and extract the portions with frequency $\omega_3 = \omega_1 + \omega_2$, thereby obtaining the generalization of Eq. (10.33):

$$\nabla^2 \left(A_i^{(3)} e^{i(k_3 z - \omega_3 t)} \right) - \frac{\partial^2}{\partial x^i \partial x^j} \left(A_j^{(3)} e^{i(k_3 z - \omega_3 t)} \right) + \frac{\omega_3^2}{c^2} \epsilon_{ij} A_j^{(3)} e^{i(k_3 z - \omega_3 t)}$$

$$= -2 \frac{\omega_3^2}{c^2} d_{ijk} A_j^{(1)} A_k^{(2)} e^{i k_3 z - \omega_3 t}. \tag{10.53}$$

(b) Explain why $e^{i(\mathbf{k}_3 \cdot \mathbf{x} - \omega_3 t)} \mathbf{f}^{(3)}$ satisfies the homogeneous wave equation (10.50). Then, splitting each wave into its scalar field and polarization vector, $A_i^{(n)} \equiv \mathcal{A}^{(n)} f_i^{(n)}$, and letting each $\mathcal{A}^{(n)}$ be a function of z (because of the boundary condition that the three-wave mixing begins at the crystal face $z = 0$), show that Eq. (10.53) reduces to

$$\alpha_3 d \mathcal{A}^{(3)} / dz = i (k_3 / \mathfrak{n}_3^2) d_{ijk} f_j^{(1)} f_k^{(2)} \mathcal{A}^{(1)} \mathcal{A}^{(2)},$$

where α_3 is a constant of order unity that depends on the relative orientations of the unit vectors $\mathbf{e}_z$, $\mathbf{f}^{(3)}$, and $\hat{\mathbf{k}}_3$. Note that, aside from α_3, this is the same evolution equation as for our idealized isotropic, dispersion-free medium, Eq. (10.34a). Show that, similarly, $\mathcal{A}^{(1)}(z)$ and $\mathcal{A}^{(2)}(z)$ satisfy the same equations (10.34b) and (10.34c) as in the idealized case, aside from multiplicative constants α_1 and α_2.

(c) Adopting the renormalizations $\mathcal{A}^{(n)} = \zeta_n \sqrt{2\omega_n / (\epsilon_0 c \, \mathfrak{n}_n)} \, \mathfrak{A}_n$ [Eq. (10.45)] with ζ_n so chosen that the photon-number flux for wave n is proportional to $|\mathfrak{A}_n|^2$, show that your evolution equations for $\mathcal{A}_n$ become Eqs. (10.48), except that the factor β' and thence the value of κ might be different for each equation.

(d) Since the evolution entails one photon with frequency ω_1 and one with frequency ω_2 annihilating to produce a photon with frequency ω_3, it must be that $d|\mathfrak{A}_1|^2 / dz = d|\mathfrak{A}_2|^2 / dz = -d|\mathfrak{A}_3|^2 / dz$. (These are called *Manley-Rowe relations*.) By imposing this on your evolution equations in part (c), deduce that all three coupling constants κ must be the same, and thence also that all three β' must be the same; therefore the evolution equations take precisely the claimed form, Eqs. (10.48).

Exercise 10.11 **Derivation: Magnitudes of Three-Wave-Mixing Quantities
Derive Eqs. (10.49). [Hint: The maximum energy flux in a wave arises from the limit $E \lesssim 10^6 \, \text{V m}^{-1}$ on the wave's electric field to ensure that it not pull electrons out of the surface of the nonlinear medium. The maximum coupling constant κ arises from the largest values $|d_{ijk}| \lesssim 10 \, \text{pm V}^{-1}$ for materials typically used in three-wave mixing (Box 10.2).]

10.7 Applications of Three-Wave Mixing: Frequency Doubling, Optical Parametric Amplification, and Squeezed Light

10.7.1 Frequency Doubling

Frequency doubling (also called *second harmonic generation*) is one of the most important applications of wave-wave mixing. As we have already discussed briefly in Secs. 10.6.2 (Track One) and 10.6.3 (Track Two), it can be achieved by passing a single wave (which plays the role of both wave $n = 1$ and wave $n = 2$) through a nonlinear crystal, with the propagation direction chosen to satisfy the resonance conditions. As we have also seen in the previous section (Track Two), the crystal's birefringence and dispersion have little influence on the growth of the output wave, $n = 3$ with $\omega_3 = 2\omega_1$; it grows with distance inside the crystal at a rate given by Eqs. (10.47), which is the same as in the Track-One case of a medium that is isotropic at linear order, Eqs. (10.38), aside from the factor β of order unity in the coupling constant κ. By doing a sufficiently good job of phase matching (satisfying the resonance conditions) and choosing the thickness of the crystal appropriately, one can achieve close to 100% conversion of the input-wave energy into frequency-doubled energy. More specifically, if wave 1 enters the crystal at $z = 0$ with $\mathfrak{A}_1(0) = \mathfrak{A}_{1o}$, which we choose (without loss of generality) to be real, and if there is no incoming wave 3 so $\mathfrak{A}_3(0) = 0$, then the solution to the evolution equations (10.47) or (10.38) is

$$\mathfrak{A}_3 = \frac{i}{\sqrt{2}}\,\mathfrak{A}_{1o} \tanh\left(\frac{\kappa}{\sqrt{2}}\,\mathfrak{A}_{1o}\,z\right), \quad \mathfrak{A}_1 = \mathfrak{A}_{1o}\,\mathrm{sech}\left(\frac{\kappa}{\sqrt{2}}\,\mathfrak{A}_{1o}\,z\right). \quad (10.54)$$

It is easy to see that this solution has the following properties. (i) It satisfies energy conservation, $2|\mathfrak{A}_3|^2 + |\mathfrak{A}_1|^2 = |\mathfrak{A}_{1o}|^2$. (ii) At a depth $z = 1.246/(\kappa\mathfrak{A}_{1o})$ in the crystal, half the initial energy has been frequency doubled. (iii) As z increases beyond this half-doubling depth, the light asymptotes to fully frequency doubled.

One might expect the frequency doubling to proceed onward to $4\omega_1$, and so forth. However, it typically does not, because these higher-frequency waves typically fail to satisfy the crystal's dispersion relation.

As an example, the neodymium:YAG (Nd^{3+}:YAG) laser, which is based on an yttrium-aluminum-garnet crystal with trivalent neodymium impurities, is among the most attractive of all lasers for a combination of high frequency stability, moderately high power, and high efficiency. However, this laser operates in the infrared, at a wavelength of 1.064 microns. For many purposes, one wants optical light. This can be achieved by frequency doubling the laser's output, thereby obtaining 0.532-micron (green) light. This is how green laser pointers, used in lecturing, work (though in 2016 they are typically driven not by Nd:YAG but rather a relative; see Ex. 10.13).

Frequency doubling also plays a key role in laser fusion, where intense, pulsed laser beams, focused on a pellet of fusion fuel, compress and heat the pellet to high densities and temperatures. Because the beam's energy flux is inversely proportional to the area of its focused cross section—and because the larger the wavelength, the more seriously diffraction impedes making the cross section small—in order to achieve efficient compression of the pellet, it is important to give the beam a very short wavelength. This is achieved by multiple frequency doublings, which can and do occur in the experimental setup of laser fusion.

EXERCISES

Exercise 10.12 *Derivation: Saturation in Frequency Doubling*
Derive the solution (10.54) to the evolution equations (10.47) for frequency doubling, and verify that it has the claimed properties.

Exercise 10.13 **Example: Frequency Doubling in a Green Laser Pointer*
Green laser pointers, popular in 2016, have the structure shown in Fig. 10.13. A battery-driven infrared diode laser puts out 808-nm light that pumps a Nd:YVO$_4$ laser crystal (neodymium-doped yttrium vanadate; a relative of Nd:YAG). The 1,064-nm light beam from this Nd:YVO$_4$ laser is frequency doubled by a KTP crystal, resulting in 532-nm green light. An infrared filter removes all the 880-nm and 1,064-nm light from the output, leaving only the green.

(a) To make the frequency doubling as efficient as possible, the light is focused to as small a beam radius ϖ_o as diffraction allows as it travels through the KTP crystal. Assuming that the crystal length is $L \simeq 3$ mm, show that $\varpi_o \simeq \sqrt{\lambda L/(4\pi \mathrm{n_1})} \simeq$

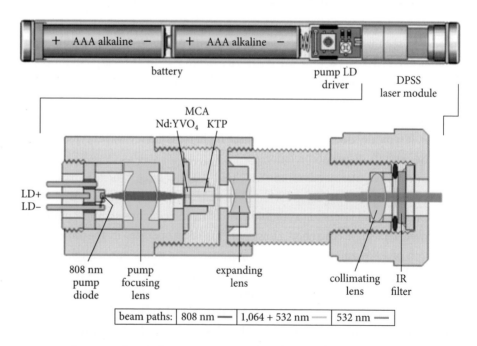

FIGURE 10.13 Structure of a green laser pointer, circa 2012. Adapted with minor changes from a drawing copyright by Samuel M. Goldwasser (Sam's Laser FAQ at http://www.repairfaq.org/lasersam .htm), and printed here with his permission.

12 μm (about 12 times larger than the 1,064-nm wavelength). [Hint: Use the properties of Gaussian beams; Sec. 8.5.5 adjusted for propagation in a medium with index of refraction n_1.]

(b) The 1,064-nm beam has an input power $W_{1o} \simeq 100$ mW as it enters the KTP crystal. Show that its energy flux and its electric field strength are $F \simeq 230$ MW m^{-2} and $\mathcal{A}^{(1)} \simeq 400$ kV m^{-1}.

(c) Assuming that phase matching has been carried out successfully (i.e., photon energy and momentum conservation have been enforced), explain why it is reasonable to expect the quantity $\beta d_{ijk} f_i^{(1)} f_j^{(1)} f_k^{(3)}$ in the coupling constant κ to be roughly 4 pm/V [cf. Eq. (6) of Box 10.2]. Then show that the green output beam at the end of the KTP crystal has $|\mathcal{A}_3|^2 \sim 0.7 \times 10^{-4} \mathcal{A}_{1o}^2$, corresponding to an output power $W_3 \sim 1.5 \times 10^{-4} W_{1o} \simeq 0.015$ mW. This is far below the output power, 5 mW, of typical green laser pointers. How do you think the output power is boosted by a factor $\sim 5/0.015 \simeq 300$?

(d) The answer is (i) to put reflective coatings on the two ends of the KTP crystal so it becomes a Fabry-Perot resonator for the 1.064 μm input field; and also (ii) make the input face (but not the output face) reflective for the 0.532 μm green light. Show that, if the 1.064 μm resonator has a finesse $\mathcal{F} \simeq 30$, then the green-light output power will be increased to $\simeq 5$ mW.

(e) Explain why this strategy makes the output power sensitive to the temperature of the KTP crystal. To minimize this sensitivity, the crystal is oriented so that its input and output faces are orthogonal to its (approximate) symmetry axis—the z-axis—for which the thermal expansion coefficient is very small (0.6×10^{-6}/C, by contrast with $\simeq 10 \times 10^{-6}$/C along other axes. Show that, in this case, a temperature increase or reduction of 6 C from the pointer's optimal 22 C (room temperature) will reduce the output power from 5 mW to much less than 1 mW. Astronomers complain that green laser pointers stop working outdoors on cool evenings.

10.7.2 Optical Parametric Amplification

optical parametric amplification; pump wave, signal wave, idler wave

In optical parametric amplification, the energy of a *pump wave* is used to amplify an initially weak *signal wave* and also amplify an uninteresting *idler wave*. The waves satisfy the resonance conditions with $\omega_p = \omega_s + \omega_i$. The pump wave and signal wave are fed into an anisotropic nonlinear crystal, propagating in (nearly) the same direction, with nonzero renormalized amplitudes $\mathcal{A}_{po}$ and $\mathcal{A}_{so}$ at $z = 0$. The idler wave has $\mathcal{A}_{io} = 0$ at the entry plane. Because the pump wave is so strong, it is negligibly influenced by the three-wave mixing (i.e., $\mathcal{A}_p$ remains constant inside the crystal).

The evolution equations for the (renormalized) signal and idler amplitudes are

$$\frac{d\mathcal{A}_s}{dz} = i\kappa \mathcal{A}_p \mathcal{A}_i^*, \qquad \frac{d\mathcal{A}_i}{dz} = i\kappa \mathcal{A}_p \mathcal{A}_s^* \qquad (10.55)$$

[Eqs. (10.48) or (10.37)]. For the initial conditions of weak signal wave and no idler wave, the solution to these equations is

wave amplitude evolution

$$\mathfrak{A}_s = \mathfrak{A}_{so} \cosh(|\gamma|z), \qquad \mathfrak{A}_i = \frac{\gamma}{|\gamma|}\mathfrak{A}_{so}^* \sinh(|\gamma|z); \qquad \gamma \equiv i\kappa\mathfrak{A}_p. \quad (10.56)$$

Thus the signal field grows exponentially, after an initial pause, with an *e*-folding length $1/|\gamma|$, which for strong three-wave nonlinearities is of order 1 cm [Ex. (10.14)].

EXERCISES

Exercise 10.14 *Derivation: e-Folding Length for an Optical Parametric Amplifier*
Estimate the magnitude of the *e*-folding length for an optical parametric amplifier that is based on a strong three-wave nonlinearity.

10.7.3

10.7.3 Degenerate Optical Parametric Amplification: Squeezed Light

Consider optical parametric amplification with the signal and idler frequencies identical, so the idler field is the same as the signal field, and the pump frequency is twice the signal frequency: $\omega_p = 2\omega_s$. This condition is called *degenerate*. Adjust the phase of the pump field so that $\gamma = i\kappa\mathfrak{A}_p$ is real and positive. Then the equation of evolution for the signal field is the same as appears in frequency doubling [Eqs. (10.47) or (10.38)]:

degenerate optical parametric amplification

$$d\mathfrak{A}_s/dz = \gamma\mathfrak{A}_s^*. \quad (10.57)$$

The resulting evolution is most clearly understood by decomposing $\mathfrak{A}_s$ into its real and imaginary parts (as we did in Ex. 6.23 when studying thermal noise in an oscillator): $\mathfrak{A}_s = X_1 + iX_2$. Then the time-evolving electric field is

$$E \propto \Re(\mathfrak{A}_s e^{i(k_s z - \omega_s t)}) = X_1 \cos(k_s z - \omega_s t) + X_2 \sin(k_s z - \omega_s t). \quad (10.58)$$

Therefore, X_1 is the amplitude of the field's cosine quadrature, and X_2 is the amplitude of its sine quadrature. Equation (10.57) then says that $dX_1/dz = \gamma X_1$, $dX_2/dz = -\gamma X_2$, so we have

$$X_1 = X_{1o}e^{\gamma z}, \qquad X_2 = X_{2o}e^{-\gamma z}. \quad (10.59)$$

Therefore, the wave's cosine quadrature gets amplified as the wave propagates, and its sine quadrature is attenuated. This is called *squeezing*, because X_2 is reduced (squeezed) while X_1 is increased. It is a phenomenon known to children who swing; see Ex. 10.15.

squeezing

Squeezing is especially interesting when it is applied to noise. Typically, a wave has equal amounts of noise in its two quadratures (i.e., the standard deviations ΔX_1 and ΔX_2 of the two quadratures are equal, as was the case in Ex. 6.23). When such a wave is squeezed, its two standard deviations get altered in such a way that their product is unchanged:

$$\boxed{\Delta X_1 = \Delta X_{1o}\, e^{\gamma z}; \qquad \Delta X_2 = \Delta X_{2o}\, e^{-\gamma z}, \qquad \Delta X_1 \Delta X_2 = \text{const}} \quad (10.60)$$

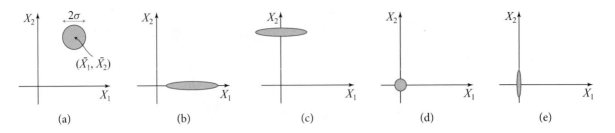

FIGURE 10.14 Error boxes in the complex amplitude plane for several different electromagnetic waves: (a) Classical light. (b) Phase-squeezed light. (c) Amplitude-squeezed light. (d) The quantum electrodynamical vacuum. (e) The squeezed vacuum.

(see Fig. 10.14). When, as here, the standard deviations of two quadratures differ, the light is said to be in a *squeezed state*.

squeezed state of light

In quantum theory, X_1 and X_2 are complementary observables; they are described by Hermitian operators that do not commute. The uncertainty principle associated with their noncommutation implies that their product $\Delta X_1 \Delta X_2$ has some minimum possible value. This minimum is achieved by the wave's vacuum state, which has $\Delta X_1 = \Delta X_2$ with values corresponding to a half quantum of energy (vacuum fluctuations) in the field mode that we are studying. When this "quantum electrodynamic vacuum" is fed into a degenerate optical parametric amplifier, the vacuum noise gets squeezed in the same manner [Eq. (10.59)] as any other noise.

squeezed vacuum

Squeezed states of light, including this *squeezed vacuum,* have great value for fundamental physics experiments and technology. Most importantly, they can be used to reduce the photon shot noise of an interferometer below the standard quantum limit of $\Delta N = \sqrt{N}$ (Poisson statistics), thereby improving the signal-to-noise ratio in certain communications devices and in laser interferometer gravitational-wave detectors such as LIGO (Caves, 1981; McClelland et al., 2011; Oelker et al., 2016). We explore some properties of squeezed light in Ex. 10.16.

EXERCISES

Exercise 10.15   **Example: Squeezing by Children Who Swing*
A child, standing in a swing, bends her knees and then straightens them twice per swing period, making the distance ℓ from the swing's support to her center of mass oscillate as $\ell = \ell_0 + \ell_1 \sin 2\omega_0 t$. Here $\omega_0 = \sqrt{g\ell_0}$ is the swing's mean angular frequency.

(a) Show that the swing's angular displacement from vertical, θ, obeys the equation of motion

$$\frac{d^2\theta}{dt^2} + \omega_0^2 \theta = -\omega_1^2 \sin(2\omega_0 t)\theta, \qquad (10.61)$$

where $\omega_1 = \sqrt{g\ell_1}$, and θ is assumed to be small, $\theta \ll 1$.

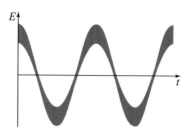

FIGURE 10.15 The error band for the electric field $E(t)$, as measured at a fixed location in space, when phase-squeezed light passes by.

(b) Write $\theta = X_1 \cos \omega_0 t + X_2 \sin \omega_0 t$. Assuming that $\ell_1 \ll \ell_0$ so $\omega_1 \ll \omega_0$, show that the child's knee bending (her "pumping" the swing) squeezes X_1 and amplifies X_2 (parametric amplification):

$$X_1(t) = X_1(0)e^{-[\omega_1^2/(2\omega_o)]t}, \qquad X_2(t) = X_2(0)e^{+[\omega_1^2/(2\omega_o)]t} \qquad (10.62)$$

(c) Explain how this squeezing is related to the child's conscious manipulation of the swing (i.e., to her strategy for increasing the swing's amplitude when she starts up, and her strategy for reducing the amplitude when she wants to quit swinging).

Exercise 10.16 **Example: Squeezed States of Light*

Consider a plane, monochromatic electromagnetic wave with angular frequency ω, whose electric field is expressed in terms of its complex amplitude $X_1 + iX$ by Eq. (10.58). Because the field (inevitably) is noisy, its quadrature amplitudes X_1 and X_2 are random processes with means $\bar{X}_1$, $\bar{X}_2$ and variances ΔX_1, ΔX_2.

(a) Normal, classical light has equal amounts of noise in its two quadratures. Explain why it can be represented by Fig. 10.14a.

(b) Explain why Fig. 10.14b represents *phase-squeezed light,* and show that its electric field as a function of time has the form shown in Fig. 10.15.

(c) Explain why Fig. 10.14c represents *amplitude-squeezed light,* and construct a diagram of its electric field as a function of time analogous to Fig. 10.15.

(d) Figure 10.14d represents the vacuum state of light's frequency-ω plane-wave mode. Give a formula for the diameter of the mode's circular error box. Construct a diagram of the electric field as a function of time analogous to Fig. 10.15.

(e) Figure 10.14e represents the squeezed vacuum. Construct a diagram of its electric field as a function of time analogous to Fig. 10.15.

10.8 Four-Wave Mixing in Isotropic Media

10.8.1 Third-Order Susceptibilities and Field Strengths

The nonlinear polarization for four-wave mixing, $P_i^{(4)} = 4\epsilon_0 \chi_{ijkl} E_j E_k E_l$ [Eq. (10.21)], is typically smaller than that, $P_i^{(3)} = 2\epsilon_0 d_{ijk} E_j E_k$, for three-wave mix-

TABLE 10.1: Materials used in four-wave mixing

Material	Wavelength (μm)	n	χ_{1111} (pm^2 V^{-2})	n_2 (10^{-20} m^2 W^{-1})
Fused silica	0.694	1.455	56	3
SF$_6$ glass	1.06	1.77	590	21
CS$_2$ liquid	1.06	1.594	6,400	290
2-methyl-4-nitroaniline				
(MNA) organic crystal[a]		1.8	1.7×10^5	5,800
PTS polydiacetylene				
polymeric crystal[a]		1.88	5.5×10^5	1.8×10^4

a. Also has large d_{ijk}.

Notes: At the indicated light wavelength, n is the index of refraction, χ_{1111} is the third-order nonlinear susceptibility, and n_2 is the Kerr coefficient of Eq. (10.69). Adapted from Yariv and Yeh (2007, Table 8.8), whose χ_{1111} is $1/\epsilon_0$ times ours.

ing by $\sim E|\chi/d| \sim (10^6 \text{ V m}^{-1})(100 \text{ pm V}^{-1}) \sim 10^{-4}$. (Here we have used the largest electric field that solids typically can support.) Therefore (as we have already discussed), only when d_{ijk} is greatly suppressed by isotropy of the nonlinear material does χ_{ijkl} and four-wave mixing become the dominant nonlinearity. And in that case, we expect the propagation lengthscale for strong, cumulative four-wave mixing to be $\sim 10^4$ larger than that ($\sim 1 \text{ cm}$) for the strongest three-wave mixing (i.e., $\ell_{4w} \gtrsim 100 \text{ m}$).

order of magnitude estimate for strength of four-wave mixing

In reality, as we shall see in Sec. 10.8.2, this estimate is overly pessimistic. In special materials, ℓ_{4w} can be less than a meter (though still much bigger than the three-wave mixing's centimeter). Two factors enable this. (i) If the nonlinear material is a fluid (e.g., CS$_2$) confined by solid walls, then it can support somewhat larger electric field strengths than a nonlinear crystal's maximum, 10^6 V m^{-1}. (ii) If the nonlinear material is made of molecules significantly larger than 10^{-10} m (e.g., organic molecules), then the molecular electric dipoles induced by an electric field can be significantly larger than our estimates (10.26); correspondingly, $|\chi_{ijkl}|$ can significantly exceed $(100 \text{ pm V}^{-1})^2$; see Table 10.1.

In Secs. 10.8.2 and 10.8.3, we give (i) an example with strong four-wave mixing: phase conjugation by a half-meter-long cell containing CS$_2$ liquid and then (ii) an example with weak but important four-wave mixing: light propagation in a multikilometer-long fused-silica optical fiber.

10.8.2 Phase Conjugation via Four-Wave Mixing in CS$_2$ Fluid

10.8.2

As an example of four-wave mixing, we discuss phase conjugation in a rectangular cell that contains carbon disulfide (CS$_2$) liquid (Fig. 10.16a). The fluid is pumped by two

phase conjugation by four-wave mixing

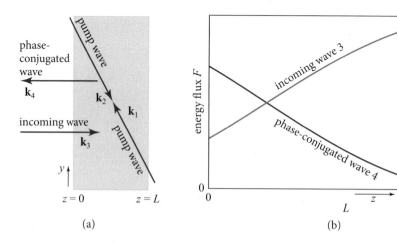

(a)   (b)

FIGURE 10.16 (a) A phase-conjugating mirror based on four-wave mixing. (b) The evolution of the incoming wave's flux and the phase-conjugated wave's flux inside the mirror (the nonlinear medium).

strong waves, 1 and 2, propagating in opposite directions with the same frequency as the incoming wave 3 that is to be phase conjugated. The pump waves are planar without modulation, but wave 3 has a spatial modulation (slow compared to the wave number) that carries, for example, a picture; $\mathcal{A}_3 = \mathcal{A}_3(x, y; z)$. As we shall see, nonlinear interaction of the two pump waves 1 and 2 and the incoming wave 3 produces outgoing wave 4, which is the phase conjugate of wave 3. All four waves propagate in planes of constant x and have their electric fields in the x direction, so the relevant component of the third-order nonlinearity is $\chi_{xxxx} = \chi_{1111}$.

resonance conditions for four-wave mixing

The resonance conditions (photon energy and momentum conservation) for this four-wave mixing process are $\omega_4 = \omega_1 + \omega_2 - \omega_3$ and $\mathbf{k}_4 = \mathbf{k}_1 + \mathbf{k}_2 - \mathbf{k}_3$. Since the three input waves all have the same frequency, $\omega_1 = \omega_2 = \omega_3 = \omega$, the output wave 4 will also have $\omega_4 = \omega$, so this is *fully degenerate four-wave mixing*. The pump waves propagate in opposite directions, so they satisfy $\mathbf{k}_1 = -\mathbf{k}_2$, whence the output wave has $\mathbf{k}_4 = -\mathbf{k}_3$. That is, it propagates in the opposite direction to the input wave 3 and has the same frequency, as it must, if it is to be (as we claim) the phase conjugate of 3.

The nonlinear polarization that generates wave 4 is

$$P_x^{(4)} = 4\epsilon_0 \chi_{1111}(E_x^{(1)} E_x^{(2)} E_x^{(3)} + E_x^{(2)} E_x^{(3)} E_x^{(1)} + \ldots).$$

There are six terms in the sum (six ways to order the three waves), so

$$P_x^{(4)} = 24\epsilon_0 \chi_{1111} E_x^{(1)} E_x^{(2)} E_x^{(3)}.$$

Inserting

$$E_x^{(n)} = \frac{1}{2}(\mathcal{A}^{(n)} e^{i(\mathbf{k}_n \cdot \mathbf{x} - \omega_n t)} + \mathcal{A}^{(n)*} e^{i(-\mathbf{k}_n + \omega_n t)})$$

into this $P_x^{(4)}$ and plucking out the relevant term for our phase-conjugation process (with the signal wave 3 phase conjugated and the pump waves not), we obtain

$$P_x^{(4)} = 3\epsilon_0 \chi_{1111} \mathcal{A}^{(1)} \mathcal{A}^{(2)} \mathcal{A}^{(3)*} e^{i(\mathbf{k}_4 \cdot \mathbf{x} - \omega_4 t)}. \tag{10.63}$$

Inserting this expression into the wave equation for wave 4 in an isotropic medium, $(\nabla^2 + \mathfrak{n}^2 \omega_4^2 / c^2)(\mathcal{A}^{(4)} e^{i(\mathbf{k}_4 \cdot \mathbf{x} - \omega_4 t)}) = -(\omega_4^2/c^2) P_x^{(4)}$ [analog of the isotropic-medium wave equation (10.33) for three-wave mixing] and making use of the fact that the lengthscale on which wave 4 changes is long compared to its wavelength, we obtain the following evolution equation for wave 4, which we augment with that for wave 3, obtained in the same way:

$$\frac{d\mathcal{A}^{(4)}}{dz} = -\frac{3ik}{\mathfrak{n}^2} \chi_{1111} \mathcal{A}^{(1)} \mathcal{A}^{(2)} \mathcal{A}^{(3)*}, \quad \frac{d\mathcal{A}^{(3)}}{dz} = -\frac{3ik}{\mathfrak{n}^2} \chi_{1111} \mathcal{A}^{(1)} \mathcal{A}^{(2)} \mathcal{A}^{(4)*}. \tag{10.64}$$

Here we have dropped subscripts from k, since all waves have the same scalar wave number, and we have used the dispersion relation $\omega/k = c/\mathfrak{n}$ common to all four waves (since they all have the same frequency). We have not written down the evolution equations for the pump waves, because in practice they are very much stronger than the incoming and phase-conjugated waves, so they change hardly at all during the evolution.

It is convenient to change normalizations of the wave fields, as in three-wave mixing, from $\mathcal{A}^{(n)}$ (the electric field) to $\mathfrak{A}_n$ (the square root of energy flux divided by frequency, $\mathfrak{A}_n = \sqrt{F_n/\omega_n}$):

$$A^{(n)} = \sqrt{\frac{2k}{\epsilon_0 \mathfrak{n}^2}} \mathfrak{A}_n = \sqrt{\frac{2\omega_n}{\epsilon_0 c \mathfrak{n}}} \mathfrak{A}_n \tag{10.65}$$

renormalized wave amplitudes

[Eq. (10.35)]. Inserting these into the evolution equations (10.64) and absorbing the constant pump-wave amplitudes into a coupling constant, we obtain

$$\boxed{\frac{d\mathfrak{A}_4}{dz} = -i\kappa \mathfrak{A}_3^*, \quad \frac{d\mathfrak{A}_3}{dz} = i\kappa \mathfrak{A}_4^*; \quad \kappa = \frac{6\omega^2}{c^2 \mathfrak{n}^2 \epsilon_0} \chi_{1111} \mathfrak{A}_1 \mathfrak{A}_2.} \tag{10.66}$$

evolution equations for fully degenerate four-wave mixing in an isotropic medium

Equations (10.66) are our final, simple equations for the evolution of the input and phase-conjugate waves in our isotropic, nonlinear medium (CS_2 liquid). Inserting the index of refraction $\mathfrak{n} = 1.594$ and nonlinear susceptibility $\chi_{1111} = 6{,}400$ (pm/V)^2 for CS_2 (Table 10.1), the angular frequency corresponding to 1.064 μm wavelength light from, say, a Nd:YAG laser, and letting both pump waves $n = 1$ and 2 have energy fluxes $F_n = \omega |\mathfrak{A}_n|^2 = 5 \times 10^{10}$ W m^{-2} (corresponding to electric field amplitudes 6.1×10^6 V m^{-1}, six times larger than good nonlinear crystals can support), we obtain for the magnitude of the coupling constant $|\kappa| = 1/0.59$ m^{-1}. Thus, the CS_2 cell of Fig. 10.16a need only be a half meter thick to produce strong phase conjugation.

For an input wave $\mathfrak{A}_{3_0}(x, y)$ at the cell's front face $z = 0$ and no input wave 4 at $z = L$, the solution to the evolution equations (10.66) is easily found to be, in the interior of the cell:

$$\mathfrak{A}_4(x, y, z) = \frac{-i\kappa}{|\kappa|} \left(\frac{\sin[|\kappa|(z - L)]}{\cos[|\kappa|L]} \right) \mathfrak{A}_{3_0}^*(x, y),$$

$$\mathfrak{A}_3(x, y, z) = \left(\frac{\cos[|\kappa|(z - L)]}{\cos[|\kappa|L]} \right) \mathfrak{A}_{3_0}(x, y). \tag{10.67}$$

The corresponding energy fluxes, $F_n = \omega|\mathfrak{A}_n|^2$ are plotted in Fig. 10.16b, for a crystal thickness $L = 1/|\kappa| = 0.59$ m. Notice that the pump waves amplify the rightward propagating input wave, so it grows from the crystal front to the crystal back. At the same time, the interaction of the input wave with the pump waves generates the leftward propagating phase-conjugated wave, which begins with zero strength at the back of the crystal and grows (when $L \sim 1/|\kappa|$) to be stronger than the input wave at the crystal front.

EXERCISES

Exercise 10.17 **Problem: Photon Creation and Annihilation in a Phase-Conjugating Mirror
Describe the creation and annihilation of photons that underlies a phase-conjugating mirror's four-wave mixing. Specifically, how many photons of each wave are created or annihilated? [Hint: See the discussion of photon creation and annihilation for three-wave mixing at the end of Sec. 10.6.1.]

Exercise 10.18 **Problem: Spontaneous Oscillation in Four-Wave Mixing
Suppose the thickness of the nonlinear medium of the text's four-wave mixing analysis is $L = \pi/(2\kappa)$, so the denominators in Eqs. (10.67) are zero. Explain the physical nature of the resulting evolution of waves 3 and 4.

Exercise 10.19 Problem: Squeezed Light Produced by Phase Conjugation
Suppose a light beam is split in two by a beam splitter. One beam is reflected off an ordinary mirror and the other off a phase-conjugating mirror. The beams are then recombined at the beam splitter. Suppose that the powers returning to the beam splitter are nearly the same; they differ by a fractional amount $\Delta P/P = \epsilon \ll 1$. Show that the recombined light is in a strongly squeezed state, and discuss how one can guarantee it is phase squeezed, or (if one prefers) amplitude squeezed.

10.8.3 Optical Kerr Effect and Four-Wave Mixing in an Optical Fiber

Suppose that an isotropic, nonlinear medium is driven by a single input plane wave polarized in the x direction, $E_x = \Re[\mathcal{A}e^{-(kz-\omega t)}]$. This input wave produces the following polarization that propagates along with itself in resonance (Ex. 10.20):

$$P_x = \epsilon_0 \chi_0 E_x + 6\epsilon_0 \chi_{1111} \overline{E^2} E_x. \tag{10.68}$$

Here the second term is due to four-wave mixing, and $\overline{E^2}$ is the time average of the square of the electric field, which can be expressed in terms of the energy flux as $\overline{E^2} = F/(\epsilon_0 nc)$. The four-wave-mixing term can be regarded as a nonlinear correction to χ_0: $\Delta\chi_0 = 6\chi_{1111}\overline{E^2} = [6\chi_{1111}/(nc\,\epsilon_0)]F$. Since the index of refraction is $n = \sqrt{1+\chi_0}$, this corresponds to a fractional change of index of refraction given by

$$\Delta n = n_2 F, \quad \text{where } n_2 = \frac{3\chi_{1111}}{n^2 c\,\epsilon_0}. \tag{10.69}$$

This nonlinear change of n is called the *optical Kerr effect,* and the coefficient n_2 is the *Kerr coefficient* and has dimensions of $1/(\text{energy flux})$, or $\text{m}^2\,\text{W}^{-1}$. Values for n_2 for several materials are listed in Table 10.1.

optical Kerr effect; Kerr coefficient

We have already briefly discussed an important application of the optical Kerr effect: the self-focusing of a laser beam, which plays a key role in mode locked lasers (Sec. 10.2.3) and also in laser fusion.

The optical Kerr effect is also important in the optical fibers used in modern communication (e.g., to carry telephone, television, and internet signals to your home). Such fibers are generally designed to support just one spatial mode of propagation: the fundamental Gaussian mode of Sec. 8.5.5 or some analog of it. Their light-carrying cores are typically made from fused silica doped with particular impurities, so their Kerr coefficients are $n_2 \simeq 3 \times 10^{-20}\,\text{m}^2\,\text{W}^{-1}$ (Table 10.1). Although the fibers are not spatially homogeneous and the wave is not planar, one can show (and it should not be surprising) that the fibers nonetheless exhibit the optical Kerr effect, with $\Delta n = n_2 F_{\text{eff}}$. Here F_{eff}, the *effective energy flux,* is the light beam's power P divided by an effective cross sectional area, $\pi\sigma_0^2$, with σ_0 the Gaussian beam's radius, defined in Eq. (8.39): $F_{\text{eff}} = P/\pi\sigma_0^2$; see Yariv and Yeh (2007, Sec. 14.1).

As a realistic indication of the importance of the optical Kerr effect in communication fibers, consider a signal beam with mean power $P = 10$ mW and a beam radius $\sigma_0 = 5\,\mu$m and thence an effective energy flux $F_{\text{eff}} = 127$ MW m^{-2}. If the wavelength is $2\pi/k = 0.694\,\mu$m, then Table 10.1 gives $n = 1.455$ and $n_2 = 3 \times 10^{-20}\,\text{m}^2\,\text{W}^{-1}$. When this beam travels a distance $L = 50$ km along the fiber, its light experiences a phase shift

$$\Delta\phi = \frac{\Delta n}{n}kL = \frac{n_2}{n}FkL \simeq 1.2 \text{ rad.} \tag{10.70}$$

A phase shift of this size or larger can cause significant problems for optical communication. For example:

1. Variations of the flux, when pulsed signals are being transmitted, cause time-varying phase shifts that modify the signals' phase evolution (*self-phase modulation*). One consequence of this is the broadening of each pulse; another is a nonlinearly induced chirping of each pulse (slightly lower frequency at beginning and higher at end).

2. Fibers generally carry many channels of communication with slightly different carrier frequencies, and the time-varying flux of one channel can modify the phase evolution of another (*cross-phase modulation*).

Various techniques have been developed to deal with these issues (see, e.g., Yariv and Yeh, 2007, Chap. 14).

optical soliton

In long optical fibers, pulse broadening due to the nonlinear optical Kerr effect can be counterbalanced by a narrowing of a pulse due to linear dispersion (dependence of group velocity on frequency). The result is an *optical soliton:* a pulse of light with a special shape that travels down the fiber without any broadening or narrowing (see, e.g., Yariv and Yeh, 2007, Sec. 14.5). In Sec. 16.3, we study in full mathematical detail this soliton phenomenon for nonlinear waves on the surface of water; in Sec. 23.6, we study it for nonlinear waves in plasmas.

EXERCISES

Exercise 10.20 *Derivation: Optical Kerr Effect*

(a) Derive Eq. (10.68) for the polarization induced in an isotropic medium by a linearly polarized electromagnetic wave.

(b) Fill in the remaining details of the derivation of Eq. (10.69) for the optical Kerr effect.

Bibliographic Note

For a lucid and detailed discussion of lasers and their applications, see Saleh and Teich (2007), and at a more advanced level, Yariv and Yeh (2007). For less detailed but clear discussions, see standard optics textbooks, such as Jenkins and White (1976), Ghatak (2010), and Hecht (2017).

For a lucid and detailed discussion of holography and its applications, see Goodman (2005). Most optics textbooks contain less detailed but clear discussions. We like Jenkins and White (1976), Brooker (2003), Sharma (2006), Ghatak (2010), and Hecht (2017).

Wave-wave mixing in nonlinear media is discussed in great detail and with many applications by Yariv and Yeh (2007). Some readers might find an earlier book by Yariv (1989) pedagogically easier; it was written when the subject was less rich, but the foundations were already in place, and it has more of a quantum mechanical focus. The fundamental concepts of wave-wave mixing and its underlying physical processes are treated especially nicely by Boyd (2008). A more elementary treatment with focus on applications is given by Saleh and Teich (2007). Among treatments in standard optics texts, we like Sharma (2006).

REFERENCES

Abbott, B. P., R. Abbott, T. D. Abbott, M. R. Abernathy, et al. (2016a). Observation of gravitational waves from a binary black hole merger. *Physical Review Letters* **116**, 061102.

Abbott, B. P., R. Abbott, T. D. Abbott, M. R. Abernathy, et al. (2016b). GW150914: The advanced LIGO detectors in the era of first discoveries. *Physical Review Letters* **116**, 131103.

Abbott, B. P., R. Abbott, R. Adhikari, P. Ajith, et al. (2009). LIGO: the Laser Interferometer Gravitational-wave Observatory. *Reports on Progress in Physics* **72**, 076901.

Arnol'd, V. I. (1992). *Catastrophe Theory*. Cham, Switzerland: Springer.

Basov, N. G., and A. M. Prokhorov (1954). First Russian ammonia maser. *Journal of Experimental and Theoretical Physics* **27**, 431–438.

Basov, N. G., and A. M. Prokhorov (1955). Possible methods for obtaining active molecules for a molecular oscillator. *Journal of Experimental and Theoretical Physics* **28**, 249–250.

Bennett, C. A. (2008). *Principles of Physical Optics*. New York: Wiley.

Berry, M. (1990). Anticipations of the geometric phase. *Physics Today* **43(12)**, 34–40.

Berry, M. V., and C. Upstill (1980). Catastrophe optics: Morphologies of caustics and their diffraction patterns. *Progress in Optics* **18**, 257–346.

Black, E. D. (2001). An introduction to Pound–Drever–Hall laser frequency stabilization. *American Journal of Physics* **69**, 79–87.

Blandford, R. D., and R. Narayan (1992). Cosmological applications of gravitational lensing. *Annual Reviews of Astronomy and Astrophysics* **30**, 311–358.

Born, M., and E. Wolf (1999). *Principles of Optics: Electromagnetic Theory of Propagation, Interference and Diffraction of Light*. Cambridge: Cambridge University Press.

Boyd, R. W. (2008). *Nonlinear Optics*. New York: Academic Press.

Brooker, G. (2003). *Modern Classical Optics*. Oxford: Oxford University Press.

Cathey, W. T. (1974). *Optical Information Processing and Holography*. New York: Wiley.

Caves, C. M. (1980). Quantum-mechanical radiation-pressure fluctuations in an interferometer. *Physical Review Letters* **45**, 75–79.

Caves, C. M. (1981). Quantum mechanical noise in an interferometer. *Physical Review D* **23**, 1693–1708.

Cundiff, S. T. (2002). Phase stabilization of ultrashort optical pulses. *Journal of Physics D* **35**, 43–59.

Cundiff, S. T., and J. Ye (2003). Colloquium: Femtosecond optical frequency combs. *Reviews of Modern Physics* **75**, 325–342.

Eddington, A. S. (1919). The total eclipse of 1919 May 29 and the influence of gravitation on light. *Observatory* **42**, 119–122.

The Event Horizon Telescope Collaboration (2019). First Event Horizon Telescope Results. I. The Shadow of the Supermassive Black Hole. *Astrophysical Journal Letters* **875**, L1–17.

Feynman, R. P. (1966). *The Character of Physical Law*. Cambridge, Mass.: MIT Press.

Feynman, R. P., R. B. Leighton, and M. Sands (1964). *The Feynman Lectures on Physics*. Reading, Mass.: Addison-Wesley. Chapter 14 epigraph reprinted with permission of Caltech.

Francon, M., and I. Willmans (1966). *Optical Interferometry*. New York: Academic Press.

Fraunhofer, J. von (1814–1815). Determination of the refractive and color-dispersing power of different types of glass, in relation to the improvement of achromatic telescopes. *Denkschriften der Königlichen Academie der Wissenschaften zu München* **5**, 193–226.

Genzel, R., F. Eisenhauer, and S. Gillessen (2010). The galactic center massive black hole and nuclear star cluster. *Reviews of Modern Physics* **82**, 3121–3195.

Ghatak, A. (2010). *Optics*. New Delhi: McGraw-Hill.

Ghez, A. M., S. Salim, N. N. Weinberg, J. R. Lu, et al. (2008). Measuring distance and properties of the Milky Way central supermassive black hole with stellar orbits. *Astrophysical Journal* **689**, 1044–1062.

Goldstein, H., C. Poole, and J. Safko (2002). *Classical Mechanics*. New York: Addison-Wesley.

Goodman, J. W. (1985). *Statistical Optics*. New York: Wiley.

Goodman, J. W. (2005). *Introduction to Fourier Optics*. Englewood, Colo.: Roberts and Company.

Goodman, J. J., R. W. Romani, R. D. Blandford, and R. Narayan (1987). The effect of caustics on scintillating radio sources. *Monthly Notices of the Royal Astronomical Society* **229**, 73–102.

Gordon, J. P., H. J. Zeiger, and C. H. Townes (1954). Molecular microwave oscillator and new hyperfine structure in the microwave spectrum of NH_3. *Physical Review* **95**, 282–284.

Gordon, J. P., H. J. Zeiger, and C. H. Townes (1955). The maser—new type of microwave amplifier, frequency standard, and spectrometer. *Physical Review* **99**, 1264–1274.

Griffiths, D. J. (2004). *Introduction to Quantum Mechanics*. Upper Saddle River, N.J.: Prentice-Hall.

Hariharan, P. (2007). *Basics of Interferometry*. New York: Academic Press.

Hecht, E. (2017). *Optics*. New York: Addison-Wesley.

Iizuka, K. (1987). *Engineering Optics*. Berlin: Springer-Verlag.

Jenkins, F. A., and H. E. White (1976). *Fundamentals of Optics*. New York: McGraw-Hill.

Klein, M. V., and T. E. Furtak (1986). *Optics*. New York: Wiley.

Kravtsov, Y. A. (2005). *Geometrical Optics in Engineering Physics*. Oxford: Alpha Science International.

LIGO Scientific Collaboration (2015). Advanced LIGO. *Classical and Quantum Gravity* **32**, 074001.

Longhurst, R. S. (1973). *Geometrical and Physical Optics*. London: Longmans.

Macintosh, B., J. R. Graham, P. Ingraham, Q. Konopacky, et al. (2014). First light of the Gemini planet imager. *Proceedings of the National Academy of Sciences* **111**, 12661–12666.

Maiman, T. H. (1960). Stimulated optical radiation in ruby. *Nature* **187**, 493–494.

Mather, J. C., E. S. Cheng, D. A. Cottingham, R. E. Eplee Jr., et al. (1994). Measurement of the cosmic microwave background spectrum by the COBE FIRAS instrument. *Astrophysical Journal* **420**, 439–444.

McClelland, D. E., N. Mavalvala, Y. Chen, and R. Schnabel (2011). Advanced interferometry, quantum optics and optomechanics in gravitational wave detectors. *Lasers and Photonics Reviews* **5**, 677–696.

Michelson, A. A., and F. G. Pease (1921). Measurement of the diameter of α Orionis with the interferometer. *Astrophysical Journal* **53**, 249–259.

Nye, J. (1999). *Natural Focusing and Fine Structure of Light*. Bristol and Philadelphia: Institute of Physics Publishing.

Oelker, T. I., T. Isogai, J. Miller, M. Tse, et al. (2016). Audio-band frequency-dependent squeezing for gravitational-wave detectors. *Physical Review Letters* **116**, 041102.

Pedrotti, F. L., L. S. Pedrotti, and L. M. Pedrotti (2007). *Introduction to Optics*. Upper Saddle River, N.J.: Pearson.

Petters, A. O., H. Levine, and J. Wambsganss (2001). *Singularity Theory and Gravitational Lensing*. Cham, Switzerland: Springer.

Poston, T., and I. Stewart (2012). *Catastrophe Theory and Its Applications*. Mineola, N.Y.: Courier Dover Publications.

Roddier, F. (1981). The effects of atmospheric turbulence in optical astronomy. *Progress in Optics* **19**, 281–376.

Saleh, B. E., and M. C. Teich (2007). *Fundamentals of Photonics*. New York: Wiley.

Saunders, P. T. (1980). *An Introduction to Catastrophe Theory*. Cambridge: Cambridge University Press.

Schneider, P., J. Ehlers, and E. Falco (1992). *Gravitational Lenses*. Berlin: Springer-Verlag.

Sharma, K. (2006). *Optics: Principles and Applications*. New York: Academic Press.

Thom, R. (1994). *Structural Stability and Morphogenesis*. Boulder: Westview Press.

Townes, C. H. (2002). *How the Laser Happened: Adventures of a Scientist*. Oxford: Oxford University Press. Chapter 10 epigraph reprinted with permission of the publisher.

Weber, J. (1953). Amplification of microwave radiation by substances not in thermal equilibrium. *IRE Transactions of the Professional Group on Electron Devices* **3**, 1–4.

Weiss, R. (1972). Electromagnetically coupled broadband gravitational antenna. *Quarterly Progress Report of the Research Laboratory of Electronics, M.I.T.,* **105**, 54–76.

Welford, W. T. (1988). *Optics*. Oxford: Oxford University Press.

Yariv, A. (1978). Phase conjugate optics and real time holography. *IEEE Journal of Quantum Electronics* **14**, 650–660.

Yariv, A. (1989). *Quantum Electronics*. New York: Wiley.

Yariv, A., and P. Yeh (2007). *Photonics: Optical Electronics in Modern Communications*. Oxford: Oxford University Press.

Young, T. (1802). On the theory of light and colours (read in 1801). *Philosophical Transactions* **92**, 34.

Zel'dovich, B. Ya., V. I. Popovichev, V. V. Ragul'skii, and F. S. Faizullov (1972). Connection between the wavefronts of the reflected and exciting light in stimulated Mandel'shtem-Brillouin scattering. *Journal of Experimental and Theoretical Physics Letters* **15**, 160–164.

NAME INDEX

Page numbers for entries in boxes are followed by "b," those for epigraphs at the beginning of a chapter by "e," those for figures by "f," and those for notes by "n."

SUBJECT INDEX

Second and third level entries are not ordered alphabetically. Instead, the most important or general entries come first, followed by less important or less general ones, with specific applications last.

Page numbers for entries in boxes are followed by "b," those for epigraphs at the beginning of a chapter by "e," those for figures by "f," for notes by "n," and for tables by "t."

gravitational lensing, 396–404
   refractive index models for, 396–397
   Fermat's principle for, 396–397
   microlensing by a point mass, 398–401
     Einstein ring, 399, 400f
     time delay in, 401
   lensing by galaxies, 401–404, 404f
gravitational waves
   dispersion relation for, 354
gravity waves on water, 353, 355f, 356
   deep water, 353, 355f, 356
Green's functions for wave diffraction, 417
   in paraxial optics, 438
group velocity, 355
guide star, for adaptive optics, 470–471
gyroscopes
   laser, 501, 502f, 520
   on Martian rover, 409

Hamilton-Jacobi equation, 362, 375
Hamilton's equations for rays in geometric optics, 361–363, 367
Hamilton's principal function, 362, 375
Hanbury Brown and Twiss intensity interferometer, 509–511
harmonic generation by nonlinear medium, 537, 545–546, 553–555
Harriet delay line, 381
Helmholtz equation, 413
Helmholtz-Kirchhoff integral for diffraction, 414, 415f
hologram, 522–531. *See also* holography
holography, 521–531
   recording hologram, 522–525, 530
   reconstructing 3-dimensional image from hologram, 525–527, 530
   secondary (phase conjugated) wave and image, 525f, 527, 535
   types of
     simple (standard) holography, 521–528
     reflection holography, 528, 530
     white-light holography, 528
     full-color holography, 528–529
     phase holography, 528
     volume holography, 528
   applications of
     holographic interferometry, 529, 529f
     holographic lenses, 529, 530–531
Hubble Space Telescope
   images from, 400, 404
   spherical aberration in, and its repair, 426–427
Huygen's model for wave propagation, 411, 417

image processing
   via paraxial Fourier optics, 436–437, 441–445
     low-pass filter: cleaning laser beam, 441
     high-pass filter, accentuating features, 441
     notch filter: removing pixellation, 441
     convolution of two images or functions, 443–444
     phase-contrast microscopy, 442–443
     transmission electron microscope, 444–445
   speckle, 470b, 472
index of refraction, 372
   numerical values, 541b–542b, 547f, 559t
   for axisymmetric optical systems, 377
   for optical elements, 483, 486–489, 497n
   for optical fiber, 374, 447, 534
   for anisotropic crystals, 546
   for Earth's atmosphere, 466b–469b
   for model of gravitational lensing, 396
instabilities in fluid flows. *See* fluid-flow instabilities
interference by division of the amplitude, 473
interference by division of the wavefront, 458
interference fringes
   for two-beam interference, 457f, 458, 458n
   for perfectly coherent waves, 459
   for waves from an extended source, 460
   fringe visibility, 460–463, 475
   in Fresnel diffraction, 419f
   near a caustic, 452f, 453
interferogram, 475, 476
interferometer
   Fabry-Perot, 490–495
   gravitational wave. *See* laser interferometer gravitational wave detector
   Michelson, 474, 475f
   Michelson stellar, 464, 465f
   Sagnac, 501–502
   radio-telescope, 479–483
   very long baseline (VLBI), 482
   intensity, 509–511
   stellar intensity, 511
interferometric gravitational wave detector. *See* laser interferometer gravitational wave detector
interferometry, multiple-beam, 483–486

James Webb Space Telescope, 427
Jupiter, 455
JVLA (Jansky Very Large Array), 480

KDP nonlinear crystal, 541b, 546–548
Kolmogorov spectrum for turbulence, 467
   for transported quantities, 467
   in Earth's atmosphere, 466b–471b
KTP nonlinear crystal, 542b, 554–555

T2 Track Two; see page xv

PART V FLUID DYNAMICS 671

PREFACE TO *MODERN CLASSICAL PHYSICS*

The study of physics (including astronomy) is one of the oldest academic enterprises. Remarkable surges in inquiry occurred in equally remarkable societies—in Greece and Egypt, in Mesopotamia, India and China—and especially in Western Europe from the late sixteenth century onward. Independent, rational inquiry flourished at the expense of ignorance, superstition, and obeisance to authority.

Physics is a constructive and progressive discipline, so these surges left behind layers of understanding derived from careful observation and experiment, organized by fundamental principles and laws that provide the foundation of the discipline today. Meanwhile the detritus of bad data and wrong ideas has washed away. The laws themselves were so general and reliable that they provided foundations for investigation far beyond the traditional frontiers of physics, and for the growth of technology.

The start of the twentieth century marked a watershed in the history of physics, when attention turned to the small and the fast. Although rightly associated with the names of Planck and Einstein, this turning point was only reached through the curiosity and industry of their many forerunners. The resulting quantum mechanics and relativity occupied physicists for much of the succeeding century and today are viewed very differently from each other. Quantum mechanics is perceived as an abrupt departure from the tacit assumptions of the past, while relativity—though no less radical conceptually—is seen as a logical continuation of the physics of Galileo, Newton, and Maxwell. There is no better illustration of this than Einstein's growing special relativity into the general theory and his famous resistance to the quantum mechanics of the 1920s, which others were developing.

This is a book about classical physics—a name intended to capture the pre-quantum scientific ideas, augmented by general relativity. Operationally, it is physics in the limit that Planck's constant $h \rightarrow 0$. Classical physics is sometimes used, pejoratively, to suggest that "classical" ideas were discarded and replaced by new principles and laws. Nothing could be further from the truth. The majority of applications of physics today are still essentially classical. This does not imply that physicists or others working in these areas are ignorant or dismissive of quantum physics. It is simply that the issues with which they are confronted are mostly addressed classically. Furthermore, classical physics has not stood still while the quantum world was being explored. In scope and in practice, it has exploded on many fronts and would now be quite unrecognizable to a Helmholtz, a Rayleigh, or a Gibbs. In this book, we have tried to emphasize these contemporary developments and applications at the expense of historical choices, and this is the reason for our seemingly oxymoronic title, *Modern Classical Physics*.

This book is ambitious in scope, but to make it bindable and portable (and so the authors could spend some time with their families), we do not develop classical mechanics, electromagnetic theory, or elementary thermodynamics. We assume the reader has already learned these topics elsewhere, perhaps as part of an undergraduate curriculum. We also assume a normal undergraduate facility with applied mathematics. This allows us to focus on those topics that are less frequently taught in undergraduate and graduate courses.

Another important exclusion is numerical methods and simulation. High-performance computing has transformed modern research and enabled investigations that were formerly hamstrung by the limitations of special functions and artificially imposed symmetries. To do justice to the range of numerical techniques that have been developed—partial differential equation solvers, finite element methods, Monte Carlo approaches, graphics, and so on—would have more than doubled the scope and size of the book. Nonetheless, because numerical evaluations are crucial for physical insight, the book includes many applications and exercises in which user-friendly numerical packages (such as Maple, Mathematica, and Matlab) can be used to produce interesting numerical results without too much effort. We hope that, via this pathway from fundamental principle to computable outcome, our book will bring readers not only physical insight but also enthusiasm for computational physics.

Classical physics as we develop it emphasizes physical phenomena on macroscopic scales: scales where the particulate natures of matter and radiation are secondary to their behavior in bulk; scales where particles' statistical—as opposed to individual—properties are important, and where matter's inherent graininess can be smoothed over.

In this book, we take a journey through spacetime and phase space; through statistical and continuum mechanics (including solids, fluids, and plasmas); and through optics and relativity, both special and general. In our journey, we seek to comprehend the fundamental laws of classical physics in their own terms, and also in relation to quantum physics. And, using carefully chosen examples, we show how the classical laws are applied to important, contemporary, twenty-first-century problems and to everyday phenomena; and we also uncover some deep relationships among the various fundamental laws and connections among the practical techniques that are used in different subfields of physics.

Geometry is a deep theme throughout this book and a very important connector. We shall see how a few geometrical considerations dictate or strongly limit the basic principles of classical physics. Geometry illuminates the character of the classical principles and also helps relate them to the corresponding principles of quantum physics. Geometrical methods can also obviate lengthy analytical calculations. Despite this, long, routine algebraic manipulations are sometimes unavoidable; in such cases, we occasionally save space by invoking modern computational symbol manipulation programs, such as Maple, Mathematica, and Matlab.

This book is the outgrowth of courses that the authors have taught at Caltech and Stanford beginning 37 years ago. Our goal was then and remains now to fill what we saw as a large hole in the traditional physics curriculum, at least in the United States:

- We believe that every masters-level or PhD physicist should be familiar with the basic concepts of all the major branches of classical physics and should have had some experience in applying them to real-world phenomena; this book is designed to facilitate this goal.

- Many physics, astronomy, and engineering graduate students in the United States and around the world use classical physics extensively in their research, and even more of them go on to careers in which classical physics is an essential component; this book is designed to expedite their efforts.

- Many professional physicists and engineers discover, in mid-career, that they need an understanding of areas of classical physics that they had not previously mastered. This book is designed to help them fill in the gaps and see the relationship to already familiar topics.

In pursuit of this goal, we seek, in this book, to *give the reader a clear understanding of the basic concepts and principles of classical physics.* We present these principles in the language of modern physics (not nineteenth-century applied mathematics), and we present them primarily for physicists—though we have tried hard to make the content interesting, useful, and accessible to a much larger community including engineers, mathematicians, chemists, biologists, and so on. As far as possible, we emphasize theory that involves general principles which extend well beyond the particular topics we use to illustrate them.

In this book, we also seek to *teach the reader how to apply the ideas of classical physics.* We do so by presenting contemporary applications from a variety of fields, such as

- fundamental physics, experimental physics, and applied physics;
- astrophysics and cosmology;
- geophysics, oceanography, and meteorology;
- biophysics and chemical physics; and

- engineering, optical science and technology, radio science and technology, and information science and technology.

Why is the range of applications so wide? Because we believe that physicists should have enough understanding of general principles to attack problems that arise in unfamiliar environments. In the modern era, a large fraction of physics students will go on to careers outside the core of fundamental physics. For such students, a broad exposure to non-core applications can be of great value. For those who wind up in the core, such an exposure is of value culturally, and also because ideas from other fields often turn out to have impact back in the core of physics. Our examples illustrate how basic concepts and problem-solving techniques are freely interchanged across disciplines.

We strongly believe that classical physics should *not* be studied in isolation from quantum mechanics and its modern applications. Our reasons are simple:

- Quantum mechanics has primacy over classical physics. Classical physics is an approximation—often excellent, sometimes poor—to quantum mechanics.

- In recent decades, many concepts and mathematical techniques developed for quantum mechanics have been imported into classical physics and there used to enlarge our classical understanding and enhance our computational capability. An example that we shall study is nonlinearly interacting plasma waves, which are best treated as quanta ("plasmons"), despite their being solutions of classical field equations.

- Ideas developed initially for classical problems are frequently adapted for application to avowedly quantum mechanical subjects; examples (not discussed in this book) are found in supersymmetric string theory and in the liquid drop model of the atomic nucleus.

Because of these intimate connections between quantum and classical physics, quantum physics appears frequently in this book.

The amount and variety of material covered in this book may seem overwhelming. If so, keep in mind the key goals of the book: to teach the fundamental concepts, which are not so extensive that they should overwhelm, and to illustrate those concepts. Our goal is not to provide a mastery of the many illustrative applications contained in the book, but rather to convey the spirit of how to apply the basic concepts of classical physics. To help students and readers who feel overwhelmed, we have labeled as "Track Two" sections that can be skipped on a first reading, or skipped entirely— but are sufficiently interesting that many readers may choose to browse or study them. Track-Two sections are labeled by the symbol **T2** . To keep Track One manageable for a one-year course, the Track-One portion of each chapter is rarely longer than 40 pages (including many pages of exercises) and is often somewhat shorter. Track One is designed for a full-year course at the first-year graduate level; that is how we have mostly used it. (Many final-year undergraduates have taken our course successfully, but rarely easily.)

The book is divided into seven parts:

I. **Foundations**—which introduces our book's powerful *geometric* point of view on the laws of physics and brings readers up to speed on some concepts and mathematical tools that we shall need. Many readers will already have mastered most or all of the material in Part I and might find that they can understand most of the rest of the book without adopting our avowedly geometric viewpoint. Nevertheless, we encourage such readers to browse Part I, at least briefly, before moving on, so as to become familiar with this viewpoint. We believe the investment will be repaid. Part I is split into two chapters, Chap. 1 on Newtonian physics and Chap. 2 on special relativity. Since nearly all of Parts II–VI is Newtonian, readers may choose to skip Chap. 2 and the occasional special relativity sections of subsequent chapters, until they are ready to launch into Part VII, General Relativity. Accordingly, Chap. 2 is labeled Track Two, though it becomes Track One when readers embark on Part VII.

II. **Statistical Physics**—including kinetic theory, statistical mechanics, statistical thermodynamics, and the theory of random processes. These subjects underlie some portions of the rest of the book, especially plasma physics and fluid mechanics.

III. **Optics**—by which we mean classical waves of all sorts: light waves, radio waves, sound waves, water waves, waves in plasmas, and gravitational waves. The major concepts we develop for dealing with all these waves include geometric optics, diffraction, interference, and nonlinear wave-wave mixing.

IV. **Elasticity**—elastic deformations, both static and dynamic, of solids. Here we develop the use of tensors to describe continuum mechanics.

V. **Fluid Dynamics**—with flows ranging from the traditional ones of air and water to more modern cosmic and biological environments. We introduce vorticity, viscosity, turbulence, boundary layers, heat transport, sound waves, shock waves, magnetohydrodynamics, and more.

VI. **Plasma Physics**—including plasmas in Earth-bound laboratories and in technological (e.g., controlled-fusion) devices, Earth's ionosphere, and cosmic environments. In addition to magnetohydrodynamics (treated in Part V), we develop two-fluid and kinetic approaches, and techniques of nonlinear plasma physics.

VII. **General Relativity**—the physics of curved spacetime. Here we show how the physical laws that we have discussed in flat spacetime are modified to account for curvature. We also explain how energy and momentum generate this curvature. These ideas are developed for their principal classical applications to neutron stars, black holes, gravitational radiation, and cosmology.

It should be possible to read and teach these parts independently, provided one is prepared to use the cross-references to access some concepts, tools, and results developed in earlier parts.

Five of the seven parts (II, III, V, VI, and VII) conclude with chapters that focus on applications where there is much current research activity and, consequently, there are many opportunities for physicists.

Exercises are a major component of this book. There are five types of exercises:

1. *Practice.* Exercises that provide practice at mathematical manipulations (e.g., of tensors).

2. *Derivation.* Exercises that fill in details of arguments skipped over in the text.

3. *Example.* Exercises that lead the reader step by step through the details of some important extension or application of the material in the text.

4. *Problem.* Exercises with few, if any, hints, in which the task of figuring out how to set up the calculation and get started on it often is as difficult as doing the calculation itself.

5. *Challenge.* Especially difficult exercises whose solution may require reading other books or articles as a foundation for getting started.

We urge readers to try working many of the exercises, especially the examples, which should be regarded as continuations of the text and which contain many of the most illuminating applications. Exercises that we regard as especially important are designated by **.

A few words on units and conventions. In this book we deal with practical matters and frequently need to have a quantitative understanding of the magnitudes of various physical quantities. This requires us to adopt a particular unit system. Physicists use both Gaussian and SI units; units that lie outside both formal systems are also commonly used in many subdisciplines. Both Gaussian and SI units provide a complete and internally consistent set for all of physics, and it is an often-debated issue as to which system is more convenient or aesthetically appealing. We will not enter this debate! One's choice of units should not matter, and a mature physicist should be able to change from one system to another with little thought. However, when learning new concepts, having to figure out "where the 2πs and 4πs go" is a genuine impediment to progress. Our solution to this problem is as follows. For each physics subfield that we study, we consistently use the set of units that seem most natural or that, we judge, constitute the majority usage by researchers in that subfield. We do not pedantically convert cm to m or vice versa at every juncture; we trust that the reader can easily make whatever translation is necessary. However, where the equations are actually different—primarily in electromagnetic theory—we occasionally provide, in brackets or footnotes, the equivalent equations in the other unit system and enough information for the reader to proceed in his or her preferred scheme.

We encourage readers to consult this book's website, http://press.princeton.edu/titles/MCP.html, for information, errata, and various resources relevant to the book.

A large number of people have influenced this book and our viewpoint on the material in it. We list many of them and express our thanks in the Acknowledgments. Many misconceptions and errors have been caught and corrected. However, in a book of this size and scope, others will remain, and for these we take full responsibility. We would be delighted to learn of these from readers and will post corrections and explanations on this book's website when we judge them to be especially important and helpful.

Above all, we are grateful for the support of our wives, Carolee and Liz—and especially for their forbearance in epochs when our enterprise seemed like a mad and vain pursuit of an unreachable goal, a pursuit that we juggled with huge numbers of other obligations, while Liz and Carolee, in the midst of their own careers, gave us the love and encouragement that were crucial in keeping us going.

ACKNOWLEDGMENTS FOR *MODERN CLASSICAL PHYSICS*

This book evolved gradually from notes written in 1980–81, through improved notes, then sparse prose, and on into text that ultimately morphed into what you see today. Over these three decades and more, courses based on our evolving notes and text were taught by us and by many of our colleagues at Caltech, Stanford, and elsewhere. From those teachers and their students, and from readers who found our evolving text on the web and dove into it, we have received an extraordinary volume of feedback,[1] and also patient correction of errors and misconceptions as well as help with translating passages that were correct but impenetrable into more lucid and accessible treatments. For all this feedback and to all who gave it, we are extremely grateful. We wish that we had kept better records; the heartfelt thanks that we offer all these colleagues, students, and readers, named and unnamed, are deeply sincere.

Teachers who taught courses based on our evolving notes and text, and gave invaluable feedback, include Professors Richard Blade, Yanbei Chen, Michael Cross, Steven Frautschi, Peter Goldreich, Steve Koonin, Christian Ott, Sterl Phinney, David Politzer, John Preskill, John Schwarz, and David Stevenson at Caltech; Professors Tom Abel, Seb Doniach, Bob Wagoner, and the late Shoucheng Zhang at Stanford; and Professor Sandor Kovacs at Washington University in St. Louis.

Our teaching assistants, who gave us invaluable feedback on the text, improvements of exercises, and insights into the difficulty of the material for the students, include Jeffrey Atwell, Nate Bode, Yu Cao, Yi-Yuh Chen, Jane Dai, Alexei Dvoretsky, Fernando Echeverria, Jiyu Feng, Eanna Flanagan, Marc Goroff, Dan Grin, Arun Gupta, Alexandr Ikriannikov, Anton Kapustin, Kihong Kim, Hee-Won Lee, Geoffrey Lovelace, Miloje Makivic, Draza Markovic, Keith Matthews, Eric Morganson, Mike Morris, Chung-Yi Mou, Rob Owen, Yi Pan, Jaemo Park, Apoorva Patel, Alexander Putilin, Shuyan Qi, Soo Jong Rey, Fintan Ryan, Bonnie Shoemaker, Paul Simeon,

---

1. Specific applications that were originated by others, to the best of our memory, are acknowledged in the text.

Hidenori Sinoda, Matthew Stevenson, Wai Mo Suen, Marcus Teague, Guodang Wang, Xinkai Wu, Huan Yang, Jimmy Yee, Piljin Yi, Chen Zheng, and perhaps others of whom we have lost track!

Among the students and readers of our notes and text, who have corresponded with us, sending important suggestions and errata, are Bram Achterberg, Mustafa Amin, Richard Anantua, Alborz Bejnood, Edward Blandford, Jonathan Blandford, Dick Bond, Phil Bucksbaum, James Camparo, Conrado Cano, U Lei Chan, Vernon Chaplin, Mina Cho, Ann Marie Cody, Sandro Commandè, Kevin Fiedler, Krzysztof Findeisen, Jeff Graham, Casey Handmer, John Hannay, Ted Jacobson, Matt Kellner, Deepak Kumar, Andrew McClung, Yuki Moon, Evan O'Connor, Jeffrey Oishi, Keith Olive, Zhen Pan, Eric Peterson, Laurence Perreault Levasseur, Rob Phillips, Vahbod Pourahmad, Andreas Reisenegger, David Reis, Pavlin Savov, Janet Scheel, Yuki Takahashi, Clifford Will, Fun Lim Yee, Yajie Yuan, and Aaron Zimmerman.

For computational advice or assistance, we thank Edward Campbell, Mark Scheel, Chris Mach, and Elizabeth Wood.

Academic support staff who were crucial to our work on this book include Christine Aguilar, JoAnn Boyd, Jennifer Formicelli, and Shirley Hampton.

The editorial and production professionals at Princeton University Press (Peter Dougherty, Karen Fortgang, Ingrid Gnerlich, Eric Henney, and Arthur Werneck) and at Princeton Editorial Associates (Peter Strupp and his freelance associates Paul Anagnostopoulos, Laurel Muller, MaryEllen Oliver, Joe Snowden, and Cyd Westmoreland) have been magnificent, helping us plan and design this book, and transforming our raw prose and primitive figures into a visually appealing volume, with sustained attention to detail, courtesy, and patience as we missed deadline after deadline.

Of course, we the authors take full responsibility for all the errors of judgment, bad choices, and mistakes that remain.

Roger Blandford thanks his many supportive colleagues at Caltech, Stanford University, and the Kavli Institute for Particle Astrophysics and Cosmology. He also acknowledges the Humboldt Foundation, the Miller Institute, the National Science Foundation, and the Simons Foundation for generous support during the completion of this book. And he also thanks the Berkeley Astronomy Department; Caltech; the Institute of Astronomy, Cambridge; and the Max Planck Institute for Astrophysics, Garching, for hospitality.

Kip Thorne is grateful to Caltech—the administration, faculty, students, and staff—for the supportive environment that made possible his work on this book, work that occupied a significant portion of his academic career.